Space-Time Computing with Temporal Neural Networks

Synthesis Lectures on Computer Architecture

Editor

Margaret Martonosi, *Princeton University*

Synthesis Lectures on Computer Architecture publishes 50- to 100-page publications on topics pertaining to the science and art of designing, analyzing, selecting, and interconnecting hardware components to create computers that meet functional, performance, and cost goals. The scope will largely follow the purview of premier computer architecture conferences, such as ISCA, HPCA, MICRO, and ASPLOS.

Quantum Computing for Computer Architects

Tzvetan S. Metodi and Frederic T. Chong

2006

Space-Time Computing with Temporal Neural Networks
James E. Smith

ISBN: 978-3-031-00626-5 print
ISBN: 978-3-031-01754-4 ebook

DOI 10.1007/979-3-031-01754-4

A Publication in the Springer series
SYNTHESIS LECTURES ON COMPUTER ARCHITECTURE #39
Series Editor: Margaret Martonosi, Princeton University

Series ISSN: 1935-3235 Print   1935-3243 Electronic

Space-Time Computing with Temporal Neural Networks

James E. Smith
Professor Emeritus, University of Wisconsin

SYNTHESIS LECTURES ON COMPUTER ARCHITECTURE #39

ABSTRACT

Understanding and implementing the brain's computational paradigm is the one true grand challenge facing computer researchers. Not only are the brain's computational capabilities far beyond those of conventional computers, its energy efficiency is truly remarkable. This book, written from the perspective of a computer designer and targeted at computer researchers, is intended to give both background and lay out a course of action for studying the brain's computational paradigm. It contains a mix of concepts and ideas drawn from computational neuroscience, combined with those of the author.

As background, relevant biological features are described in terms of their computational and communication properties. The brain's neocortex is constructed of massively interconnected neurons that compute and communicate via voltage spikes, and a strong argument can be made that precise spike timing is an essential element of the paradigm. Drawing from the biological features, a mathematics-based computational paradigm is constructed. The key feature is spiking neurons that perform communication and processing in space-time, with emphasis on time. In these paradigms, time is used as a freely available resource for both communication and computation.

Neuron models are first discussed in general, and one is chosen for detailed development. Using the model, single-neuron computation is first explored. Neuron inputs are encoded as spike patterns, and the neuron is trained to identify input pattern similarities. Individual neurons are building blocks for constructing larger ensembles, referred to as "columns". These columns are trained in an unsupervised manner and operate collectively to perform the basic cognitive function of pattern clustering. Similar input patterns are mapped to a much smaller set of similar output patterns, thereby dividing the input patterns into identifiable clusters. Larger cognitive systems are formed by combining columns into a hierarchical architecture. These higher level architectures are the subject of ongoing study, and progress to date is described in detail in later chapters. Simulation plays a major role in model development, and the simulation infrastructure developed by the author is described.

KEYWORDS

spiking neural networks, temporal models, unsupervised learning, classification, neuron models, computing theory

Contents

Figure Credits

Preface 2019

Since the time of this book's publication, the author's approach to developing Temporal Neural Networks (TNNs) has continued to evolve. In general, this evolution has been in the direction of increasing simplicity rather than increasing complexity. The following paragraphs summarize the most significant changes.

The biggest change in approach has been with respect to synaptic modeling and training. Both in Part III and in the lead-up material in Part II, emphasis is placed on compound synapses. A compound synapse is the composition of multiple simple synapses that connect the same two neurons. Each connection has a different delay. Compound synapses are both biologically plausible and computationally very expressive. This also makes them difficult to work with when developing a new computing paradigm. Consequently, the author has begun working with simpler synapse neurons in order to better understand the detailed operation of STDP before proceeding to compound synapses.

On a related matter, the averaging approach to synaptic training has been abandoned in favor of a more conventional Spike Timing Dependent Plasticity (STDP) approach, wherein the excitatory column, lateral inhibitory column, and STDP work closely together in a coordinated way, using only information local to each synapse and its associated neuron body. The averaging method was intended to simplify and streamline the simulation process. However, this approach does not generate synaptic weights that are similar to weights produced by conventional STDP. Unlike compound synapses which may be an avenue for productive future research, the averaging technique for training does not appear to be viable.

A more conventional STDP approach is also advantageous because it is naturally amenable to continual learning, in which both evaluation (inference) and training are intertwined ongoing processes. In the long run, this feature may prove to be one of the most important aspects of TNNs. Localized STDP is an essential element of the emergent learning behavior that will be crucial to the eventual success of this enterprise, and devising efficient, robust, localized STDP is a hard problem.

With regard to training inhibitory blocks, the author no longer uses the Pareto optimizing approach used in Section 7.7. Although this method may eventually prove to be useful for some types of TNNs, it is does not appear to be necessary for the TNNs under consideration here. Currently, inhibition parameters are manually specified, and the same parameters hold for all inhibitory columns in the same network layer.

With regard to input encoding, biologically plausible "OnOff" encoding computes the difference between a center pixel and the average of its surround. At the time this book was originally written in 2017, it was felt that similar computational properties could be achieved with an encoding composed of both the positive and negative of an image (Section 7.3). However, a closer approximation to true OnOff encoding has been found to work better. Furthermore, OnOff encoding is relatively easy to compute, so there are no apparent advantages to the positive-negative approach.

Space-time algebra and its association with Generalized Race Logic (GRL) were added late in the development of the book. This aspect of the work appears to be very promising, and a fuller development can be found in the paper: James E. Smith (2018). "Space-time algebra: A model for neocortical computation." In *Proceedings of the 45th Annual International Symposium on Computer Architecture*, pp. 289–300, DOI: 10.1109/ISCA.2018.00033.

With regard to new areas for TNN research that are not discussed in the book, dendritic computation provides significant potential for innovation. With dendritic computation, input spikes coming into the same dendrite can interact in ways that implement simple functions. For example, in terms of space-time algebra, max and min functions may be implemented in the dendrites, prior to STDP. This opens up the possibility of operations that are akin to pooling operations in conventional machine learning systems.

J. E. Smith

April 30, 2019

Preface 2017

Understanding, and then replicating, the basic computing paradigms at work in the brain will be a monumental breakthrough in both computer engineering and theoretical neuroscience. I believe that the breakthrough (or a series of breakthroughs) will occur in the next 40 years, perhaps significantly sooner. This means that it will happen during the professional lifetime of anyone beginning a career in computer engineering today. For some perspective, consider the advances in computation that can be accomplished in 40 years.

When I started working in computer architecture and hardware design in 1972, the highest performing computer was still being constructed from discrete transistors and resistors. The CDC 7600, which began shipping in 1969, had a clock frequency that was two orders of magnitude slower than today's computers, and it had approximately 512 KB of main memory. Now, 40+ years later, we have clock frequencies measured in GHz and main memories in GB. The last 40 years has been an era of incredible technology advances, primarily in silicon, coupled with major engineering and architecture refinements that exploit the silicon advances.

It is even more interesting to consider the 40 years prior to 1972. That was a period of fundamental invention. In the early 1930s, Church, Gödel, and Turing were just coming onto the scene. Less than 30 years prior to 1972, in 1945, von Neumann wrote his famous report. Twenty years prior, in 1952, Seymour Cray was a newly minted engineer. Just before our 1972 division point, the CDC 7600 had a very fast in-order instruction pipeline and used a RISC instruction set (although the term hadn't yet been coined). Also in the first 40 years, cache memory, micro-coding, and issuing instructions out-of-order had already been implemented in commercially available computers; so had virtual memory and multi-threading.

To summarize: the most recent 40 years of computer architecture, spanning an entire career, has largely been a period of application and technology-driven *refinement*. In contrast, the 40 years before that was an era of great *invention*—the time when the giants walked the earth. Based on an admittedly self-taught understanding of neuroscience, I believe we are at the threshold of another 40 years of great invention—inventing an entirely new class of computing paradigms.

Computer architects and engineers have a number of roles to play in the discovery of new computing paradigms as used in the brain's neocortex. One role is developing large scale, Big Data platforms to manage and analyze all the information that will be generated by connectome-related projects. Another role is developing special purpose computing machines to support high performance and/or efficient implementations of models proposed by theoretical neuroscientists.

The role that I emphasize, however, is as an active participant in formulating the underlying theory of computation. That is, computer architects and engineers should be actively engaged in proposing, testing, and improving plausible computational methods that are fundamentally similar to those found in the neocortex.

Computer architecture and engineering, in the broad sense, encompasses CMOS circuits, logic design, computer organization, instruction set architecture, system software, and application software. Someone knowledgeable in computer architecture and engineering has a significant understanding of the entire spectrum of computing technologies from physical hardware to high-level software. This is a perspective that no other research specialization has.

I am sure that many computer architects and engineers would love to work on the extremely challenging, far-reaching problem of understanding and implementing paradigms as done in the brain. Unfortunately, there is a significant barrier to entry. That barrier is the daunting mountain of neuroscience literature combined with the well-established scientific culture that has grown up alongside it (e.g., vocabulary, representation style, mathematical style). This isn't insignificant, by the way: the language you use shapes the way you think.

So, how to overcome this barrier? Answering that question is the over-arching theme of this book.

First, it requires a lot of reading from the mountain of neuroscience literature; there is no shortcut, but the papers cited herein can provide some guidance. Then, by taking a bottom-up approach, computer architects and engineers will achieve their best leverage. At the bottom is the abstraction from biological neural systems to a mathematics-based formulation. In conventional computers, the analogous abstraction is from CMOS circuits to Boolean algebra. A computer architect, with some perspective and insight, can start with biological circuits (as complicated as they may seem) and model/abstract them to a practical mathematics-based computational method.

This book contains a description of relevant biological features as background. Then drawing from these biological features, a mathematics-based computational paradigm is constructed. The key feature is spiking neurons that perform communication and processing in *space-time*, with emphasis on *time*. In these paradigms, time is used as a freely available resource for communication and computation. Along the way, a prototype architecture for implementing feedforward data clustering is developed and evaluated.

Although a number of new ideas are described, many of the concepts and ideas in this book are not original with the author. Lots of ideas have been proposed and explored over the decades. At this point, there is much to be gained by carefully choosing from existing concepts and ideas, then combining them in new and interesting ways—engineering, in other words.

The particular modeling choices made in this book are but one set of possibilities. It is not even clear that the methods explored in this book are eventually going to work as planned. At the time of this writing, the ongoing design study in the penultimate chapter ends with an incomplete prototype neural network design.

There is no doubt many other approaches and modeling choices that could be, and should be, explored. Eventually, someone will find just the right combination of ideas—and there is every reason to expect that person will be a computer engineer.

***

If the reader's goal is to achieve a solid understanding of spike-based *Neural Computation*, then this book alone is not enough. It is important to read material from the literature concurrently. There is a long list of references at the end of this book; too long to be a practical supplementary reading list. Following is a shorter, annotated list of background material. None of the listed papers is standalone; rather, each provides a good touchstone for a particular area of related research.

Neuron-Level Biology:

Just about any introductory textbook will do. Better yet, use a search engine to find any of a number of excellent online articles, many illustrated with nice graphics.

Circuit-Level Biology:

Mountcastle, Vernon B. "The columnar organization of the neocortex." *Brain* 120, no. 4 (1997): 701–722.

Buxhoeveden, Daniel P. and Manuel F. Casanova. "The minicolumn hypothesis in neuroscience." *Brain* 125, no. 5 (2002): 935–951.

Hill, Sean L., et al. "Statistical connectivity provides a sufficient foundation for specific functional connectivity in neocortical neural microcircuits." *Proceedings of the National Academy of Sciences* (2012): E2885–E2894.

The paper by Mountcastle is a classic, mostly summarizing his groundbreaking work on the column hypothesis. The paper by Buxhoeveden and Casanova is an excellent review article. The paper by Hill et al. is from the Markram group in Switzerland; it is experimental work that attempts to answer the right kinds of questions regarding connectivity.

Modeling:

Morrison, Abigail, Markus Diesmann, and Wulfram Gerstner. "Phenomenological models of synaptic plasticity based on spike timing." *Biological Cybernetics* 98, no. 6 (2008): 459–478.

The operation of synapses is critical to the computational paradigm, and this is an excellent modeling paper specifically directed at synapses and synaptic plasticity. This and other work by Gerstner and group should be at the top of any reading list on neuron modeling.

Theory:

Maass, Wolfgang. "Computing with spiking neurons." In *Pulsed Neural Networks*, W. Maass and C. M. Bishop, editors, pages 55, 85. MIT Press (Cambridge), 1999.

> *Maass did some seminal theoretical research in spiking neural networks. This paper summarizes much of that work along with related research by others.*

Computation:

Masquelier, T., and Simon J. Thorpe. "Unsupervised learning of visual features through spike timing dependent plasticity." *PLoS Computational Biology* 3, no. 2 (2007): e31.

Bohte, Sander M., et al., "Unsupervised clustering with spiking neurons by sparse temporal coding and multilayer RBF networks." *IEEE Transactions on Neural Networks*, 13, no. 2 (2002): 426–435.

Karnani, Mahesh, et al. "A blanket of inhibition: Functional inferences from dense inhibitory connectivity." *Current Opinion in Neurobiology* 26 (2014): 96–102.

> *Simon Thorpe is a pioneer in spiking neural networks of the type described in this book. All the work by Thorpe and associates is interesting reading, not just the paper listed here. The work by Bohte et al., builds on earlier radial basis function research, which should also be read. The paper by Karnani et al. is a nice discussion of inhibition and the modeling thereof.*

Machine Learning:

Ciresan, Dan, et al. "Flexible, high performance convolutional neural networks for image classification." *Proceedings of the Twenty-Second International Joint Conference on Artificial Intelligence.* 2 (2011): 1237–1242.

> *The neural networks being developed in this book fit within the broad discipline of machine learning. Consequently, there are some similarities with conventional machine learning approaches. This paper describes a deep convolutional neural network of about the same scale as the networks studied here.*

Meta-Theory:

Chaitin, Gregory. *META MATH! The Quest for Omega.* Vintage, 2008.

> *The book by Chaitin is fairly easy-to-read and is imbued with the concepts and philosophy behind algorithmic information theory. When studying a computing paradigm that is not human-designed, it is good to have a "meta-" perspective.*

Acknowledgments

I would like to thank Raquel for her love and great patience. Without her unwavering support, writing this book would not have been possible. Five years ago, Mikko Lipasti started me down this research path, along with Atif Hashmi and Andy Nere, and for that I will always be thankful. I am grateful to Mark Hill for providing the initial impetus for this book and Margaret Martonosi, his successor as editor of this series, for her insightful suggestions which had a major impact on the book's eventual direction. I thank Mario Nemirovsky for providing very helpful advice along the way, and Ravi Nair, Tim Sherwood, Abhishek Bhattacharjee, and an anonymous reviewer for their many helpful comments and suggestions. Thanks to Deb Gabriel and the staff at Morgan & Claypool for their efforts. Mike Morgan deserves a special thanks not only for his support of this book, but also for providing a forum for authors to express and disseminate their ideas.

PART I

Introduction to Space-Time Computing and Temporal Neural Networks

CHAPTER 1

Introduction

Einstein rides a bicycle—a concise summary of the brain's capabilities. It can take streams of input from the vision system, the sense of balance, and the sense of place. It can combine and process this information to generate an orchestrated plan that drives myriad motor neurons to fire in a precisely coordinated way to keep the bicycle upright and moving in a chosen direction. And then there is Albert Einstein. The brain can reason deeply and abstractly, divorced from all the senses. It can conceptualize the cosmos as well as things happening deep inside an atom.

Photo courtesy of the Archives, California Institute of Technology.

In mammals, cognitive skills are performed in regions of the brain's *neocortex*. The neocortex is the folded gray surface layer of the brain—the part of the brain that is clearly visible in drawings or photographs. The human neocortex weighs about one kilogram, has a volume of about a liter, and consumes about 20 watts of power. Yet, it has capabilities that far exceed those of our conventional computing methods (including state-of-the-art artificial intelligence).

Despite intense study by neuroscientists for well over 100 years, most of the secrets regarding cognitive functions are still well hidden. It is heartening, however, to consider that after scientists have unraveled many of evolution's inventions, we see cleverness and simplicity. Consider the eye and the inter-operation of the cornea, the iris, and the retina. Consider the cardiovascular system.

Although composed of many cell types, operating in complex ways, controlled by a wide variety of hormones, the cardiovascular system is a marvel of architectural simplicity when separated from the implementation details. Decades from now, when the cognitive functions are finally understood, will we be stuck with painfully complicated models? Or, in retrospect, will we be amazed by the simple elegance? This book is predicated on an affirmative answer to the second question.

It is proposed that we approach the topic of neocortex-like cognition (computation) in a different way than we normally do when dealing with computational methods. The starting point of an alternative approach is to consider systems in which time is an essential communication and computation resource. As will be explained later, a computing system based on such a *space-time* model is unlike almost every computational system we have constructed. A good reason that evolution may have selected such an approach is that time has some ultimate engineering advantages: it is freely available, consumes no power, and requires no space.

Consequently, the objective of this book is to propose and explore computational paradigms that may belong to the same class as those employed in the neocortex. This exploration is done via the experimental design of computing systems that exhibit basic cognitive capabilities. Experiments take the form of detailed simulations, with simulation results feeding back into the paradigm formulation process.

Before going any further, we start with some essential background summarizing the way neurons work.

1.1   BASICS OF NEURON OPERATION

According to the more-than-century-old neuron doctrine [12, 105], the *neuron* is the fundamental element for structure and function in the brain's neocortex. Huge numbers of interconnected neurons communicate and compute via voltage pulses or *spikes*. When viewed as a computing problem, then, the principal computational element is the neuron, and voltage spikes are the primary means of encoding and communicating information.

Figure 1.1 illustrates two of the vast number of interconnected neurons that make up the neocortex. A neuron is a specialized cell that consists of a *body* surrounded by a *membrane*. The neuron's body is fed through inputs—*dendrites*, and it has an output—the *axon*, which branches out and is connected to the dendrites of many other neurons. A *synapse* is at the connection point between an axon and a dendrite.

Figure 1.1 shows the electrical behavior of biological neurons, *excitatory* neurons in this example. In the figure, a synapse is the connection point between an upstream neuron and a downstream neuron. A single neuron has thousands of such synapses attached to its dendrites, connecting it to hundreds or thousands of axons driven by upstream neurons.

In Figure 1.1, three virtual probes (dashed arrow lines) measure voltage levels at certain points in this tiny neural circuit. Consider a sequence of events that begins with a spike being emitted from the upstream neuron. The spike travels along the axon and reaches the synapse connecting it with the downstream neuron. Upon arriving at the synapse, the spike opens a conductive gate via a relatively complex biochemical process. The now-open conductive gate immediately allows ions to flow into the downstream neuron body, thereby raising its membrane potential. The effect on membrane potential is shown as the *response* waveform near the bottom of the figure. A more detailed picture of the relationship between a spike and its response is in Figure 1.2.

A synapse has an associated *weight*, which controls its relative conductivity. A stronger synapse has higher conductivity, resulting in a response function with higher amplitude. To continue the gate metaphor, the greater the synaptic weight, the wider open the ion gate swings in response to an incoming voltage spike.

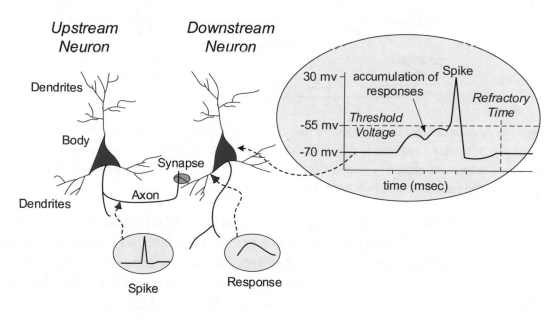

Figure 1.1: Neuron operation. Two (excitatory) pyramidal neurons are connected via a synapse. Attached dotted line "probes" illustrate dynamic operation.

Immediately after the synapse's conductive gate swings open, it starts to close with exponential decay, so the flow of ions into the downstream neuron gradually diminishes. All the while, ions leak from the neuron body, thereby decreasing the membrane potential also with exponential decay, but with a longer time constant than closing the conductive synapse gates. This combination of exponential decays with different time constants gives the excitatory response its distinctive shape, as shown in Figure 1.2. It is loosely analogous to filling a leaky bucket (the neuron body), from another leaky bucket (the synapse), where the lower body bucket leaks more slowly than the upper synapse bucket. The neuron's membrane potential is the amount of water in the body bucket.

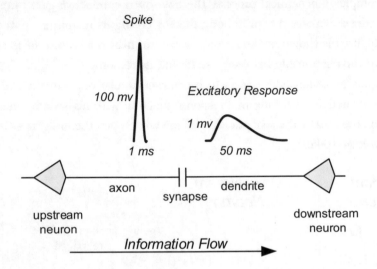

Figure 1.2: A spike emitted by an excitatory neuron flows along out along the axon. When it reaches the synapse, it invokes a response in the membrane potential of the downstream neuron (the excitatory response). The response caused by a spike from an inhibitory neuron is similar, except the polarity is reversed and the time constants tend to be faster.

Next, consider the more detailed waveform shown on the right side of Figure 1.1. As multiple spikes are received at a neuron's input synapses, each of them will contribute its excitatory response to the neuron body. If received relatively closely together in time, the effects of the responses accumulate, raising the total membrane potential, as shown in the waveform. If the potential reaches a critical level, the *threshold voltage*, then an avalanche effect takes place, and the neuron emits an output spike that is transmitted along its axon to all of its downstream neurons. Immediately following the output spike, there is a brief *refractory period*, during which the neuron cannot fire.

In some instances, there may be insufficient input spikes to raise the membrane potential to the threshold voltage. Consequently, the membrane potential will eventually decay back to the rest potential and no spike will be emitted.

In the waveform at the right side of Figure 1.1, typical voltage levels are shown. These are a base level of roughly -70 mv, a threshold of about -55 mv, and a maximum spike level of about 30 mv. The refractory time is on the order of a few msec. In the illustration, the accumulated responses of only three input spikes in close proximity are sufficient to raise the membrane potential to the threshold. In reality, the number of spikes is commonly an order of magnitude higher, but can vary over a wide range.

As just described, synaptic weights are a major factor in determining a given neuron's overall function. Weights are established via a training process, where commonly occurring spike patterns are learned and encoded as patterns of synaptic weights. Consequently, when such a learned pattern is detected at the neuron's input, the synaptic weights yield a strong response, and the neuron emits an output spike. The closer the pattern matches the learned pattern as reflected in the weights, the sooner the output spike occurs.

The commonly used rationale for synaptic weigh training is that if the upstream neuron's spike *precedes* the output spiking of the downstream neuron, it contributes to the occurrence of the downstream neuron's spike, so its influence should be enhanced in the future by increasing the synaptic weight. On the other hand, if the upstream neuron's spike occurs *after* the downstream neuron's spike, it had nothing to do with the downstream neuron's spike so its influence should be diminished in the future (by decreasing its synaptic weight). Eventually, all the weights converge to values that collectively characterize the input spike pattern.

There are two basic types of neurons: *excitatory* and *inhibitory*. Roughly 80% of neurons are excitatory, and most of them are pyramidal neurons, as exemplified in Figure 1.1. A spike from an inhibitory neuron invokes an *inhibitory response* when it reaches the synapses. An inhibitory response has the opposite polarity of an excitatory response so it reduces the downstream neuron's membrane potential. In this book, excitatory neurons are modeled as just described. However, inhibitory neurons are modeled at a higher level.

The spike propagation delay from one neuron to the next is of the same order as the processing latency of a neuron (the time between input and output spikes), and there can be fairly wide variations in both propagation delay and processing latency. Furthermore, when one neuron feeds another, there are often multiple connection paths between the two. That is, the axon from the upstream neuron makes multiple connections with dendrites belonging to the same downstream neuron. The propagation delays along these paths may differ; in fact, there is enough irregularity in dendrite and axon structure that the multiple propagation delays almost certainly differ.

Terms

The above description of neuron operation uses common terms for some of the neuron features and functions, although some also have technical terms. In this book, common terms are preferred and are used whenever possible. For reference, following is the correspondence between common terms and technical terms, along with their associated acronyms.

Common Term	More Technical Term
body	soma
spike	action potential (AP)
response	post synaptic potential (PSP)
excitatory response	excitatory post synaptic potential (EPSP)
inhibitory response	inhibitory post synaptic potential (IPSP)
upstream neuron	pre-synaptic neuron
downstream neuron	post-synaptic neuron

Schematics and Symbols

It is often useful to draw schematic diagrams of interconnected neurons to illustrate their inter-operation. In the neuroscience literature there doesn't seem to be a widely accepted standard for schematics, and schematics are usually drawn in a way that reflects the relative physical orientation and location of biological neurons.

In this book, schematics are drawn in a way that makes functional relationships apparent, without consideration for such things as physical orientation. Signal (spike train) flow is usually from left to right. The neuron symbols used throughout this book are illustrated in Figure 1.3.

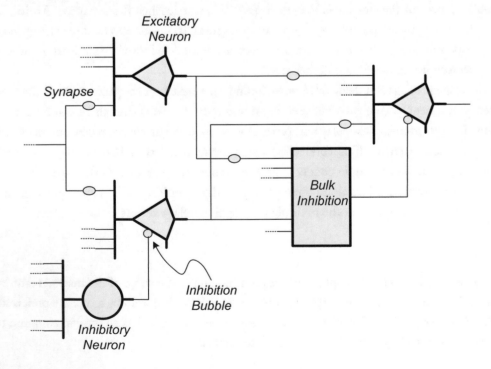

Figure 1.3: Illustration of symbols used in figures.

An excitatory neuron has the general shape of a pyramid, turned on its side. Inputs are on the left and the output is on the right. An inhibitory neuron is drawn as a circle. In this work, however, inhibitory neurons will seldom be drawn individually. Rather, because inhibitory neurons operate in bulk, a collection of inhibitory neurons will be drawn as a single box. Inhibition will then be represented at the target sites (excitatory neurons) as a bubble.

The symbol for a synapse is optional and usually not used; it only reflects the presence of a connection, which is redundant when the connection is explicitly drawn. Nevertheless, for illustrative purposes, it is sometimes useful to explicitly draw a synapse.

1.2   SPACE-TIME COMMUNICATION AND COMPUTATION

Any physical computer consumes both space and time—it operates in *space-time*. The concept of space-time is usually employed in modern physics to describe non-intuitive relativistic effects: the warping of space-time near a black hole. However, space-time also exists in a boring non-warped form almost everywhere else. We live and breathe in a uniform space-time environment where the passage of time can be a useful communication and computation resource.

Here we use the physics term "space-time" rather than more commonly used "spatio-temporal". There are two reasons for this. First, the two-syllable "space-time" rolls off the tongue much more easily than the five-syllable "spatio-temporal." Second, and more importantly, we are interested in computing methods that rely on the passage of physical time.

1.2.1   COMMUNICATION

The potential of the "time" part of space-time is illustrated by the simple system in Figure 1.4. Information is transmitted from sender to receiver over a channel; call it a *wire*, but the communication medium could be anything, an axon for example. Information is conveyed through voltage pulses or *spikes* that travel on the wire having a *constant* transmission delay ($\Delta 0$).

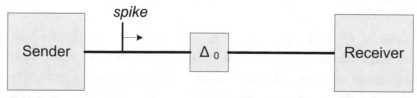

Figure 1.4: A single wire system for communicating information. Spikes travel from sender to receiver with a fixed delay.

For sake of discussion, assume both the sender and receiver have the ability to measure time intervals to within 1 msec. Hence, the sender is able to emit spikes that are separated by time intervals to a precision of 1 msec, and the receiver is able to determine the time interval between two received spikes, also to a precision of 1 msec. These measurements are performed independently at both the sender and receiver. As will now be shown, this capability yields an extremely energy efficient communication method based on spatially separated time measurements.

Assume the total energy cost for communication comes from generating, transmitting, and receiving individual spikes. Then a simple energy efficiency metric is the ratio of information sent (in bits) to the number of spikes that transmit the information. If n spikes transmit b bits of information, then the *energy efficiency metric* is b/n bits per spike. A higher number is better.

Although the term "spike" is used, spikes can be given conventional binary interpretations. In terms of bits, the communication model is one in which 0's and 1's are transmitted over the wire at 1 msec intervals. Very importantly, sending a binary 1 (a spike) consumes energy, but sending a 0 (non-spike) does not.

In order to send a packet of information over the wire, one could use conventional bit-serial encoding. That is, the sender can first send a synchronizing *start* spike on the wire, then immediately at 1 msec intervals it can send a serial binary sequence of spikes/non-spikes representing 1's/0's. For example, to send a byte of information, there is a start spike plus a bit-serial pattern containing from 0 to 8 spikes spread over eight 1 msec intervals. If all byte patterns are equally likely to be sent over the wire, then an average of 1 + 4 spikes will be required for sending a byte of information. This is an average energy efficiency of 8 bits per 5 spikes or 1.6 bits per spike.

An alternative method, is to send a *start* spike, wait n msec, then send a *stop* spike, where n is a binary number between 1 and 256: essentially the content of the byte. The information "n" is encoded by an n msec time interval between *start* and *stop* spikes. Because the sender and receiver have timing agreement to within 1 msec (due to uniform space-time), both the sender and receiver will agree on the intervening time interval. In this scenario 8 bits of information are transmitted by 2 spikes, so the energy efficiency is 4 bits per spike. This is over twice as efficient as the 1.6 bits per spike provided by the bit-serial scheme. Furthermore, if a bus is used instead of a single wire and all the wires in the bus share the same start spike, it is relatively easy to approach 8 bits per spike on a bus of modest width.

Where does the high energy efficiency come from? The answer follows from the observation that the bit-serial scheme takes a total of 9 msec to communicate a byte, and the two-spike scheme takes an average of about 128 msec. The two methods illustrate a tradeoff between energy consumption and time. By taking more time, we use less energy. From the perspective of evolution developing the neocortex, the "time" part of space-time may well have been used in a similar manner as an effective way of reducing energy requirements. In the neocortex, data precision is very low, three to four bits—not even a byte. This means that communication times are reasonable, even if not as good as bit-serial. Furthermore, increased communication latencies can be at least partially offset by the vast parallelism in the neocortex.

Observe that with this approach, the time required to transmit information *is* the information. Furthermore, this communication method is completely self-synchronized—the mere presence of the start spike is all that is needed to coordinate the communication. So, the same voltage spikes simultaneously coordinate and communicate information. This is another, less direct way of saving energy.

1.2.2   COMPUTATION

Normally, we don't think of sending information on a wire as performing a mathematical function, but if we use time as a resource, computation can be performed simultaneously with transmitting information. Consider Figure 1.5, where there are two wires with different delays: one is Δ_0 as before, and the other is Δ_0 plus 1 time unit (say 1 msec). Here, the communication process performs the elementary function $f(n) = n + 1$. The time difference between the receipt of the spike on the first wire and the receipt of the spike on the second wire determines the function's output. It is as if the sender pushes an encoded value n into a function evaluation box and $n + 1$ pops out the other side.

In general, by selecting the delay appropriately, one can both communicate and perform $f(n) = n + c$ for any constant c. Observe that the time required to compute the sum *is* the sum.

Although $n + c$ is a simple function, evaluating it consumes energy in a conventional digital CMOS implementation. That is, when adding two binary numbers, several energy-burning signal transitions are required per adder stage. In the case of the wire system, however, addition is a byproduct of communication over paths having different delays. The only cost is transmitting two spikes over wires with fixed delays—the addition is performed simultaneously with communication.

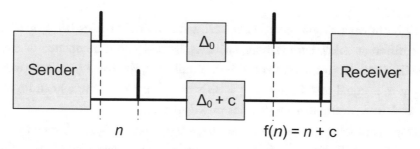

Figure 1.5: Spike-based computation of $f(n) = n + 1$.

In the neocortex this apparently simple method of computation-through-delay potentially leads to a fairly complex set of controllable computations. Recall from the neuron overview that two neurons are typically connected via several different paths with different delays. Consequently, the synapses on these different paths, via their trained weights, may selectively activate and/or de-activate the multiple parallel paths in some coordinated way, thereby achieving a programmable family of functions—which are implemented through delays in the communications paths. This is illustrated in Figure 1.6.

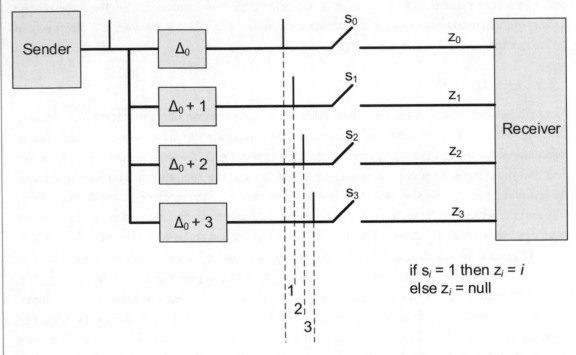

Figure 1.6: Sending a single spike results in multiple received spikes having different delays. Switches (synapses) then control the presence (closed switch) or absence (open switch) of a received value.

Figure 1.6 shows a single spike being sent, with multiple copies being received and each copy having a different delay. Furthermore, switches, labeled s_i, can be opened or closed to allow the delayed spike through to the receiver or not. Simply put, for every i, if the switch s_i is closed, then the output z_i is equal to i. In this way, a fairly complex vector (the z_i) can be constructed at the receiving end, in response to a single transmitted spike.

For example, in Figure 1.6 if $s_0 = 1$ (closed), $s_1 = 0$ (open), $s_2 = 1$, $s_3 = 1$, then the output is $z_0 = 0$, $z_1 = $ null, $z_2 = 2$, $z_3 = 3$. This may seem like an odd little function, but depending on the switch settings, a large set of output vectors can be produced in response to an input spike. That is, the switch settings (synapses) can encode a rather rich amount of information with a single transmitted spike.

The wire examples just given implement their functions in a passive way. Active, more complex computation can be performed by devices that take spike times as an input encoding and then emit an output spike at a time that is a function of the input spike times. Reflecting back to the neuron overview given above, this is exactly what a biological neuron does. That is, each input spike produces a response function of time, the response functions from multiple input spikes are added, and when (if) a threshold is met, an output spike is generated.

There is plenty of experimental support for active computation based on spike timing in the neocortex. For example, excitatory neurons are capable of maintaining a precision of 1 msec: "stimuli with fluctuations resembling synaptic activity produced spike trains with timing reproducible to less than 1 millisecond" [60]. This and other experimental support for computation via precise neuron timing will be laid out in Section 3.5.6.

1.2.3   DISCUSSION

Using time as a communication and computation resource has at least two significant advantages: energy efficiency and the ability to compute via localized self-timed elements (neurons) while achieving a coordinated global outcome; coordination is provided by the uniform passage of physical time. For these reasons, and others, a fundamental hypothesis motivating the work in this book is that the neocortex uses time in this way.

Hypothesis: The neocortex uses time as an important communication and computation resource. Results are entirely dependent on the physical time it takes to compute and communicate them.

Although a physical implementation that relies on time as a freely available resource may have significant implementation advantages, striving for a neuromorphic hardware implementation that literally uses time in this way is not an objective of the work presented here. In this work, we are not aiming for highly efficient analog networks that literally communicate via voltage spikes.

Rather, the premise is that advantages of using time as a communication/computation resource pushed evolution in a certain direction: toward a particular way of computing in the neocortex, toward a certain class of computational paradigms. And, very importantly, these paradigms are fundamentally unlike the ones we normally use when designing conventional computer systems.

It is asserted that the way that we (computer designers) ordinarily deal with time restricts the class of functions we can implement in a natural, intuitive way. Consequently, we tend to implement the class of functions that are readily amenable to Turing machine or von Neumann style implementations based on algorithmic steps, where the physical time consumed in implementing each step does not affect values computed during that step. With all computation models that we normally use, results computed during a time step are independent of the time consumed in a physical implementation.

It is further asserted that implementations that rely on physical time as a computing resource readily support a different class of functions than those supported by time-independent implementations.

If the above assertions are true, then the neocortex supports functions that are different than the ones supported by our conventional time-independent computer implementations. Consequently, the types of computation models used in the neocortex may be concealed from us because we innately think only about computing systems with time-independent implementations.

The issue is not whether such cognitive functions are theoretically computable in the Turing sense, because they are. Rather, the issue is whether our way of thinking about computation, and the role of time in particular, inhibits us from *implementing* them, or possibly even considering them. In contrast, driven by advantages such as simplicity and low energy consumption, evolution may have discovered effective time dependent methods for implementing cognitive functions.

So, at a high level, it is the general class of space-time computing paradigms we are interested in. However, because they differ from the way we ordinarily think about computing, we can't use our ordinary methods to describe and develop them. If we can't use our familiar methods, how do we proceed? The answer is that we put forward some general principles, then use experimentally observed operation of the biological neocortex for guidance. We then explore a broad new class of paradigms, from the bottom up, in a trial-and-error fashion, starting with the example that we know exists ("I think, therefore I am"). However, we are not restricted to biologically based examples—there may be other ways of using space-time besides the way the neocortex uses it. It is hoped that exploration of space-time methods, starting with neocortex-like operation, will eventually reveal a new class of computational methods that are capable of greatly expanded cognitive functions.

A more formal definition of space-time computing and its connection to operation of the neocortex are given in Chapter 2. Because the work in this book falls under the broad categories of both neural networks and machine learning, the following two background sections provide some perspective.

1.3   BACKGROUND: NEURAL NETWORK MODELS

The computing systems considered in this book are a certain type of spiking neural network (SNN). The next section discusses what this means and places the particular type of SNNs studied here into the larger context of artificial neural networks (ANNs). A simple taxonomy of neural networks is developed, based on (1) the ways that information is encoded and (2) the ways that neurons operate on this information.

The assumption is that interneuron communication is via streams of spikes (*spike trains*) generated by upstream neurons and transmitted over axons. An input spike train passes through synapses and dendrites and causes downstream neurons to respond as described in the preceding section.

1.3.1   RATE CODING

The classical way of interpreting the information contained in spike trains is to use spike rates. In the taxonomy, there are two ways of representing rates; both are shown in Figure 1.7a. One way represents information in original spike train form, i.e., individual spikes are tracked in the neural network. The other rate coding representation encodes information as numerical values. Both methods are based on the assumption that spike rates on individual lines encode information independently of the rate codings on other lines. This is illustrated in Figure 1.7b, where a change in the third line from 8 to 4 has no effect on the other lines.

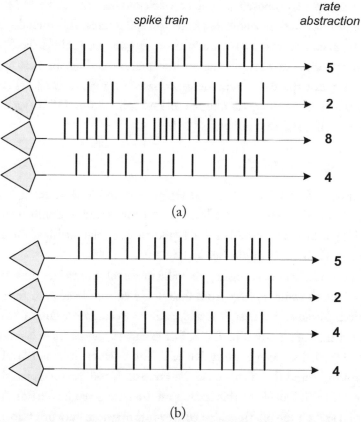

Figure 1.7: Information encoded as spike rates. (a) Spike rate information may be represented as a spike train, or as an abstract, numerical, rate. (b) If the spikes on one of the lines changes (the third line is reduced from 8 to 4), the rates (information) on the other lines does not change.

In Figure 1.7, rates are normalized so they fit within the range 0 to 8. Classical ANNs fit this coding model; in classical ANNs rates are typically scaled to the range of 0 to 1. Also included are the more recent convolutional neural networks, deep neural networks, restricted Boltzmann machines (RBMs), and various related species. These networks are the basic building blocks, the workhorses, of most of the leading-edge machine learning methods. Some may add other functional blocks, "pooling," for example, where the outputs of a number of neurons are reduced to a single value by taking the maximum or a mean [86].

Most of today's neural networks simply encode input values as numbers, and the word "rate" doesn't usually enter the conversation. However, these networks are descended from early neural networks that were based on rate assumptions, so it's in their DNA. Put it this way: if one wanted to somehow establish biological plausibility of conventional neural networks, then the plausibility argument would almost certainly rest on a numerical spike rate abstraction.

In a smaller group of proposed neural network models, the spike train representation is retained, even though information is encoded based on spike rates. This includes spike train representations based on various forms of rate modulation. A good example is the Spaun system [24]. Spaun is based on coordinated rate-modulated information coding. In that system, a group of neurons represents a vector space. A neural network based on memristors [82] is another example of a system that employs rate coding via rate modulated spike trains. GPU systems also have been applied to rate-based spiking neural networks [28].

1.3.2 TEMPORAL CODING

With temporal coding, information is encoded via precise timing relationships among individual spikes across multiple parallel transmission lines (axons connected to dendrites). There are several methods for encoding information based on such spike timing relationships. The basic method used in this book is explained via an example. See Figure 1.8.

This method encodes values according to spike times relative to the time of the first spike in a group. In Figure 1.8a, the values range from 0 to 8. Given that time goes from left to right, the leftmost spike is first. Because it is first, it is assigned the maximum value of 8. Then all the other spikes encode values that are relative to the 8. For example, the latest occurring spike (2nd line from the top) is assigned the much lower value of 2. The bottom spike is 1/3 of the relative time separating the 2 and the 8 and therefore encodes a value of 4, and the top spike that occurs slightly earlier encodes a value of 5. Details of this coding method are given in Section 6.2.

Because actual spike time relationships convey information, network transmission delays are a crucial consideration in neural networks using this coding method. A change in network delay along some axon-synapse-dendrite path can modify the information being conveyed on that path. For example, in Figure 1.8b the spike on the third line is delayed to a later time. As a consequence, not just the value on the third line changes, but *all* the values change because the spike at $t = 0$ changes. This is quite unlike what happens with rate coding when we change the value on one of the lines as described above. Here, the values on lines other than the third appear to change even though there is no apparent physical connection. In a physical implementation, however, there actually *is* a connection: physical time.

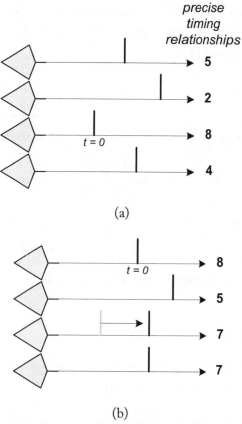

(a)

(b)

Figure 1.8: (a) Relative spike times encode information across a bundle of lines. The earliest spike has the highest value and later spikes have values that are temporally relative to the first. (b) If the spike on one of the lines is delayed (the spike on the third line shifts later in time) then the values on all the other lines change as well as the value on the third line.

1.3.3   RATE PROCESSING

In networks where information is coded as abstract rates (numbers), the neurons naturally operate on rates, i.e., numerical values falling in some fixed range. Abstract rate coding combined with rate-based neurons defines the classic ANNs and their many descendants.

In rate-based systems, values are confined to a certain fixed range. Given rate-coded inputs, the typical ANN paradigm first computes a vector inner product (weights · input values) followed by a non-linear activation, or "squashing," function of the resulting inner product. Refer to Figure 1.9. That is, it squashes the output value so that it fits within the pre-defined range. A simple squashing function that saturates for values outside the range is a good example. ANNs are typically constructed with a differentiable squashing function, so tanh or $1/(1-e^{-x})$ are often used. These are examples of "sigmoid" functions, so-called because of the general S-shape of their curves. Differentiability is important for training methods that employ back propagation and gradient descent based on partial derivatives [52].

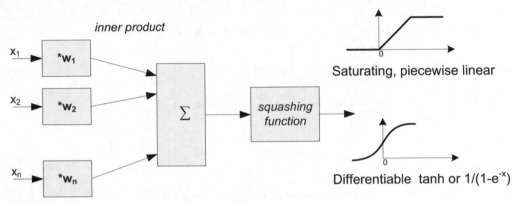

Figure 1.9: Model neuron as used in conventional neural networks.

1.3.4   SPIKE PROCESSING

A neural network that employs spike processing uses neurons that operate on spike inputs. The input spikes can either encode a rate (via a spike train) or encode a temporal vector via a single spike on each neuron input.

Regardless of the coding method, these neurons typically model neuron operation as depicted in Section 1.1, i.e., biology-like spike-and-response behavior. Such neuron models are generally called "leaky integrate and fire" (LIF) neurons. (All of this is discussed in detail in Chapter 5.)

For example, the Spaun system [24] uses leaky-integrate-and-fire neurons to operate on rate-coded spike trains. The neurons were co-developed with the rate-modulated coding system. Consequently, the spiking neurons in Spaun are able to retain a rate-modulated coding discipline throughout the system. Other systems convert input values to Poisson distributed spike bursts, with the mean determined by the input value [5, 9, 23, 76, 82]. LIF neurons then operate on these rate modulated spike trains.

1.3.5   SUMMARY AND TAXONOMY

Based on the above discussion, we can construct a neural network taxonomy that accounts for the way information is communicated and processed. Refer to Figure 1.10. On the left branch of the taxonomy, both rate-coding and rate-processing are used by all classical ANNs and their descendants, including deep convolutional neural networks.

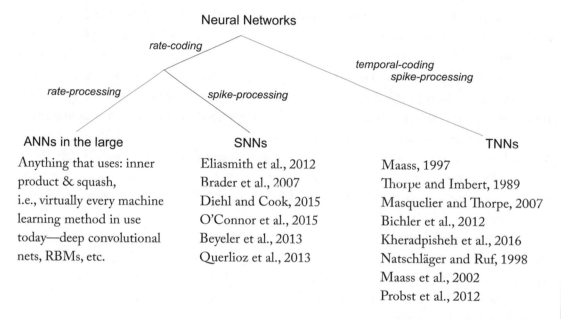

Figure 1.10: Neural network taxonomy, based on coding and processing methods. Examples are given for each. The class of networks in this book, TNNs, use both spike coding and spike processing.

The other two neural network categories, the ones that use spike processing, are generally considered to be SNNs. One type of SNN keeps rate information in spike train form and processes at the spike level. The other type of SNN, as used in this book, encodes information using precise spike timing relationships and processes information at the spike level. These SNNs of the second type are both spike-coding and spike-processing. To avoid confusion, this second type of SNN is given the specific name *Temporal Neural Network* (TNN) to emphasize the use of both temporal coding and temporal processing. Introducing a new name for these networks is not gratuitous—some of the better-known SNNs are based on rate coding; the term TNN distinguishes the methods used here.

TNNs define a broad space-time computing universe. In this universe, communication and computation delays aren't an unfortunate byproduct of a physical implementation—rather, delays and latencies are an integral part of the way communication and computation is done. A computed result is not independent of the time it takes for signals to move through a neural network. Rather, a computed result *is* the time it takes for the signals to move through the network.

1.4   BACKGROUND: MACHINE LEARNING

The applications of interest in this book and the applications targeted by conventional machine learning approaches overlap significantly, if not completely. In the broad sense of the term, the work presented here *is* machine learning. However, mainstream machine learning is not based on a space-time paradigm as considered here. On the contrary, mainstream machine learning uses time independent methods, as does virtually every other computing method that has been devised.

Nevertheless, at a high level, features of some mainstream machine learning implementations are similar, or analogous, to the features of the TNNs described in this book. One important feature of many machine-learning methods is the use of layers of interconnected feedforward functional blocks. Included in this class of systems are deep convolutional networks as well as virtually all the other feedforward neural network variants. Hence, many machine-learning approaches share the same general structure as the systems studied here: a layered hierarchy of interconnected computational units.

Both machine learning and the approach given here rely on training (or "learning," depending on the point of view). However, it is with training that a big distinction arises. Whereas with conventional machine learning, the training process is global and extremely time-consuming, the space-time approach leads to a simple, localized training process that consumes no more computation time than evaluation.

Finally, and very importantly, widely used machine learning functions—feedforward pattern clustering and classification—are excellent drivers for the initial development of TNNs. Feedforward clustering (via unsupervised training) is of primary interest in this book. Grouping according to similarity, as done with clustering, is a basic cognitive function, and there is fairly strong experimental support for biological equivalents.

In simple terms, the clustering function maps each of a large number of input patterns onto a much smaller set of output patterns based on pattern similarity. Similar patterns belong to the same cluster. Hence, the output pattern identifies a cluster, and similar input patterns map to the same cluster identifier. Typically, the number of clusters is much smaller than the size of the input space. The clustering function is determined via a training process during which input patterns to be separated into clusters are applied to the network in an unsupervised manner.

1.5   APPROACH: INTERACTION OF COMPUTER ENGINEERING AND NEUROSCIENCE

The idea of using space-time paradigms where time is a communication and computation resource is intriguing on its own, and such paradigms could be studied independently of neuroscience. However, it is the functional capabilities of the neocortex that makes space-time computing so appealing. Therefore, it makes sense to use knowledge gleaned from neuroscience to provide guidance when proposing and exploring space-time paradigms. Consequently, the search for computational paradigms with the functionality of the neocortex will involve a combination of the disciplines of neuroscience and computer engineering.

This might suggest the coordination of computer engineering and neuroscience to co-develop a single, shared theory. However, a significant problem with such a shared theory is that neuroscience and computer engineering use theory in different ways.

A theory is useful to a neuroscientist to explain and motivate experiments that expand knowledge of the biological brain. In terms of the standard computer architecture hierarchy extending from hardware implementation up through high level software, this approach is downward-looking—toward the physical "hardware," i.e., the biology. Neuroscience drives toward theory that is inclusive: include as many biological features as possible into the theory.

Theory, in the form of a computational model, is useful to a computer engineer to drive the construction of computer systems having improved and expanded capabilities. In terms of the architecture stack, this is upward-looking. Computer engineering drives toward theory (models) that are exclusive: use abstraction and modeling to exclude unnecessary implementation details.

Here is an example. In neuroscience research, once a computational model has been developed it is common to add noise and jitter to the model and show that it still works. That is, the noise and jitter naturally present in the biological system are deemed to be an important modeling consideration, and it is therefore important to show that the proposed computing model still works despite their degrading effects. In computer engineering research, on the other hand, there is no reason to artificially add noise and jitter to a model, unless they *improve* the model in some way. For a computer engineer, the thing to do with a new computing model is to design a computer.

Even within the realm of computer engineering, however, research methods aimed at cognitive processing will differ from those used by computer engineers today. With conventional computer engineering, system design generally proceeds top-down, where a desired function is given in a well-specified manner, then software and/or hardware are designed, typically using layers of abstraction.

In the construction of space-time paradigms, we don't start with a well-defined function, and we don't have known layers of abstraction that we can fall back on. Rather, we start with a general description of what a desired function should do. For example, drawing from machine learning, we might train a spiking neural network showing it a number of example faces and then ask it to identify similar faces. Importantly, we do not provide a precise functional definition of "similarity" *a priori*. Rather, the implementation and the training process imply what "similarity" means, and we can only measure the quality of the internal similarity function indirectly via benchmarking with some assumed accuracy metric.

The research process then proceeds bottom-up rather than top-down. We design a system having a certain structure, then simulate it via benchmarks, which yield an accuracy (or some similar metric), and then, based on this result, we might tweak, or perhaps significantly change, the structure (which changes the implied function) and try again. The researcher's insight and ingenuity guides the design process toward a useful computational method.

In summary, a proposed engineering research method for studying space-time computation consists of the following steps.

1. Postulate primitive communication and computation elements upon which a paradigm can be constructed. These primitive elements may be determined through the study of neuroscience research results, both theoretical and experimental.

2. Propose a computational paradigm based on the primitives. Again, the basic features may be determined via the study of neuroscience research results. There should be enough detail that a working implementation can be constructed. Implementation includes simulation models.

3. Experimentation is performed via simulation. A typical experiment consists of first training the network, then evaluating it to determine the extent to which desired cognitive capabilities are achieved.

4. Depending on the experimental results, a proposed paradigm may be accepted, modified, or rejected. Then a new, possibly expanded paradigm is put forward, possibly with a different set of primitive elements, and the process continues.

In (3) above, the experiments are directed at understanding the functionality that can be achieved. This differs from today's computer engineering simulations where the function being performed by a computer is not at issue, only the performance or energy consumption are of concern.

In (1) and (2), one might begin with features that appear *plausible* in the biological neocortex. However, in this context, the standard of plausibility may fall significantly short of standards of acceptance used by the biological neuroscience community. From the computer engineering perspective, there is no reason to be a stickler for plausibility when the ultimate goal is an innovative way of computing. One should err on the side of expansiveness. Interesting ideas should be pursued.

Neuroscience, in contrast, generally maintains a very high standard of plausibility for acceptance. Consequently, there may be a significant gap between what is accepted as plausible when trying to engineer new computing system vs. what is plausible for neuroscientific acceptance. As research progresses in both computer engineering and neuroscience, this plausibility gap should close. Ideally, over time, the gap might close completely, resulting in a unified theory that is embraced by both computer engineering and neuroscience.

1.6   BOTTOM-UP ANALYSIS: A GUIDING ANALOGY

Imagine that a civilization of space aliens, one that uses completely different computing methods than we, visits earth and discovers our state-of-the-art computer systems containing superscalar cores.

A superscalar processor contains many features to enhance performance, as well as features to improve reliability and energy efficiency. Instructions are fetched based on predicted branches; logical registers are renamed to physical registers and placed in issue buffers. Instructions are executed in parallel, often out of their original sequence. Hierarchical cache memories support both instruction fetches and data loads and stores. Clock gating enables functional units only when needed. Error detecting and correcting codes provide protection against bit failures in memories.

If the aliens try to understand the basic computing paradigm, it would be a very long, arduous task. And it would almost certainly proceed in a bottom-up fashion. The first breakthrough would probably be the realization that precise voltage levels are not important; that voltage levels can be modeled as ones and zeros. Then, after some effort, it might be discovered that logic gates perform elementary operations and that information flows from the output of one to the inputs of others.

When knowledge reaches the point of understanding combinational functions and binary arithmetic, it would be seen as a major breakthrough. Given enough components, one could then build huge combinational networks capable of performing very useful computations. A few hundred million transistors can implement an extremely large feedforward dataflow graph; obviously not with all the capabilities of a modern day processor, but extremely useful nevertheless.

Eventually, it might be discovered that the paradigm is based on synchronized feedback-connected combinational blocks. And, finally, the von Neumann paradigm might be discovered (or re-invented). Or, the aliens may arrive at a different paradigm, a dynamic dataflow architecture, for example, which is also composed of synchronized combinational blocks.

If we compare the basic von Neumann paradigm, which is quite simple, with a modern day superscalar processor the difference is huge. A simple von Neumann processor can be constructed with a few thousand logic gates, yet a superscalar processor uses several orders of magnitude more. Why the difference?

The vast majority of the logic in a modern day superscalar processor is there to increase performance, enhance reliability, and improve energy efficiency. Meanwhile, the von Neumann paradigm is concealed, deeply enmeshed in all the circuitry implementing the performance and reliability improvements. For example, the program counter is an essential part of the von Neumann architecture. Yet, if one were to try to find *the* program counter in a superscalar processor, the task is ill-stated. There isn't a single program counter; there may be many tens or even hundreds, depending on the number of concurrent in-flight instructions. A branch predictor would probably drive the aliens crazy. It consumes a lot of transistors, yet computed results do not depend on it. Whatever its predictions, the computed results are always the same.

It may be the same in the neocortex, which is a mass of interconnected neurons that communicate through an even more massive set of synapses. In essence, the neocortex is a gigantic analog system built of unreliable parts, connected via paths with variable delays. Maintaining stability of this huge analog system very likely requires constant adjustment via a complex set of biological mechanisms, as well as consuming a large fraction of neurons for maintaining reliability. Cognition itself may appear almost as a second order effect that is deeply intertwined in the multitude of supporting neurons and mechanisms. The important point here is that many of the mechanisms at work in the neocortex are very likely there for reasons other than implementing the computational paradigm. This is expanded upon in Section 4.7.

The analogy of the space aliens' bottom-up discovery of conventional computing paradigms provides some guidance in the search for the cognitive paradigm(s) used in the neocortex. There is broad consensus that voltage spikes convey information. There is also consensus regarding the basic operation of neurons. The critical next step is to understand the paradigm at a level analogous to combinational networks (i.e., feedforward networks where the output is a function of only the current input(s)). Given that understanding, one could begin to engineer large feedforward spiking neural networks capable of useful tasks. Then, armed with feedforward functionality, engineers may be able to design and construct more efficient and practical recurrent networks with feedback, thereby achieving the next big step in implementing higher-level cognitive tasks.

1.7   OVERVIEW

This book traces the author's ongoing development of a particular TNN model. It is one of many possible TNNs in the universe of space-time paradigms. A TNN based on biological neuron behavior is best developed in parallel with a simulation model. Hence, simulation is central to this work. Benchmarks, taken from the discipline of machine learning, provide simulator datasets. The overall modeling and implementation approach is illustrated in Figure 1.11.

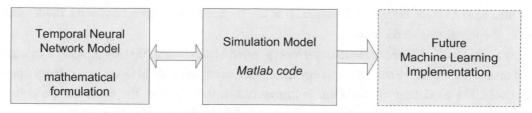

Figure 1.11: Overview of modeling and implementation methods.

A *Temporal Neural Network Model* is developed. The model specifies the way data is communicated and the way excitatory neurons and bulk inhibition behave. It also specifies the grouping of neurons into computational units that form the building blocks for scalable, hierarchical systems.

A *Simulation Model* is developed concurrently with the neural network model. The simulator implements algorithms, in high-level language software, that translate the neural network model into executable code.

Finally, the simulator is developed with an eye toward an eventual *Machine Learning Implementation*. For example, arithmetic precision during simulation is limited both for plausibility reasons and to assure feasibility of an eventual short precision fixed-point implementation. Hence, the simulation model may serve as a prototype for an eventual TNN-based machine learning implementation.

Chapter 2 is a theoretical discussion of space-time computing that provides perspective and guidance for constructing neocortex-like computing systems. A broad class of space-time computing networks is formally defined and are shown to be isomorphic to TNNs.

Chapter 3 is a biological overview that describes those features of the biological brain (the neocortex, specifically) that underlie the models developed in later chapters. It is not intended to be a summary or tutorial of what is a very broad topic; rather, it focuses on selected experimental results that provide guidance and support for the remainder of the book.

Chapter 4 begins with the experimental results from Chapter 3 and specifies an initial set of modeling assumptions that underlie the TNNs to be constructed later.

In Chapter 5, individual neuron modeling is discussed at some length. First is the development of a plausible spiking neuron model, in which each neuron operates within its own temporal frame of reference. Multiple synaptic paths connect the same pair of neurons. Because the individual paths exhibit a range of delays, they must each be included in the model. This is a critical first step toward a method of space-time computation.

Chapter 6 is a transition. It describes single excitatory neurons from a computational perspective. There are two levels of excitatory neuron development. At the lower level, individual synaptic paths are modeled. At the higher level, the multiple paths connecting two neurons are combined into a single, compound synapse. It is the higher-level compound synapse model that is used in the remainder of the book.

In Chapter 7, excitatory neurons are incorporated into columns, thereby forming a computational unit. Useful space-time computation takes place when multiple neurons in a column operate collectively. The modeling of inhibition is introduced in this chapter. The focus here, and in the remainder of this work, is on feedforward pattern clustering (unsupervised learning).

Chapter 8 describes the simulation model that is used for design space exploration. The structure of the simulator closely tracks the model definition given in the preceding chapters.

Chapter 9 describes the ongoing development of a prototype clustering TNN using the dataset from the widely used machine learning benchmark MNIST.

Chapter 10 summarizes and draws overall conclusions.

CHAPTER 2

Space-Time Computing

"Aspects of time are often said to be more 'abstract' than their spatial analogues because we can perceive the spatial, but can only imagine the temporal." [17]

Our overarching objective is to understand the types of computing paradigms employed in the neocortex, and then implement them with conventional digital technology. If there ever was a problem that calls for "thinking outside the box," this is it. Fortunately, this may be one of those rare situations where the boundaries of "the box" can actually be specified.

The box contains the ways that we ordinarily think about computing in space and time. When we reason about computation, we put most of the complexity in the spatial dimension and try to simplify the temporal dimension as much as possible. Consequently, we use only simple forms of abstract time, and these simplifying abstractions are used so universally that we seldom think about them. However, with space-time computing, time and temporal abstractions come to the forefront.

In this chapter, the argument is made that space-time paradigms may offer new ways of thinking about and designing computers with advanced cognitive capabilities. Before proceeding, some discussion of meta-architecture terms follows; these will be used throughout this chapter and the remainder of the book.

2.1 DEFINITION OF TERMS

- A *function* is a mapping from inputs to outputs, without regard to any implementation.

- A *model implementation* is a well-defined, clear, and unambiguous method for evaluating a function, i.e., a method for calculating an output for a given input. Sometimes, a model implementation is expressed in mathematical terms. An example of a model implementation is a formally defined Turing machine. Another example of a model implementation is the VHDL description of a microprocessor.

- A *physical implementation* (or *physical computer*) is a physically constructed computer system that evaluates a function. Normally, a physical implementation is based on a model implementation. An example physical implementation is a *finite-tape* Turing machine, sitting on a table, moving the physical tape to and fro (it's been done). Another example is a silicon microprocessor embedded in a smart phone. Physical time is naturally present in any physical implementation.

When a physical computer operates in a *time independent* manner, computation proceeds as a series of steps and, in a properly working physical system, the physical time for performing each step has no effect on the results. Removing the effects of physical time from each computing step is something we strive for when we build both synchronous and race-free, delay-independent asynchronous systems [44, 101].

For example, a physical finite-tape Turing machine would compute by following a sequence of steps, and the times (e.g. in seconds) required for performing a step: moving the read/write head, reading, writing, erasing, etc., has no effect on the outcome of that step. Whether the write head moves fast or slow, it always writes the same prescribed symbol. The same applies to higher-level paradigms. We write a program as a series of statements (steps), and a statement's hardware execution time (in nanoseconds, for example) has no bearing on the results of that statement. Similarly, with machine code, the physical time required to execute each machine instruction does not affect the results of the instruction. When computational theorists discuss the "time" complexity of some algorithm, they don't mean physical time. They mean the number of abstract steps, each of which can consume an unspecified amount of physical time. Many cognitive models, including high-level models as well as artificial neural network models, don't account for the passage of physical time in the implementation (and don't model it).

Not all physical computing systems are based on time independence, however. In time-dependent implementations the passage of time in the physical implementation does determine the computed results, either partially or entirely. Classical analog computers are time-dependent. In an analog computer, some varying physical quantity (e.g., voltage) naturally traces out a function of physical time. The analog computer is typically programmed by starting with a set of coupled integro-differential equations, and then implementing analogous functions (of time) with interconnected hardware blocks. Consequently, the time required by the physical device to perform an analog computation is an essential part of the analog computer's output.

So it may be with the brain. It is argued in this book that in the neocortex, the physical computation time affects the results. Hence, the neocortex may be considered an analog computer[1] of a special type. What makes it special is that rather than encoding information as waveforms tracing time-varying levels, information is encoded only by denoting points in time (with voltage spikes).

2.2   FEEDFORWARD COMPUTING NETWORKS

The neocortex is a physical computing device that consumes finite space and requires finite time for any of its computations. With that as a given, it further assumed that any function computed by the neocortex is Turing computable. Observe that Turing computability is implicit in virtually

---

1   This stretches the definition of "analog" somewhat because the temporal signals aren't really analogous to anything; "analog" is used here as the opposite of "digital."

all theoretical neuroscience. Anyone who constructs and runs a computer model for the operation of the neocortex is making this assumption.

It then follows that any of the cognitive functions performed by the neocortex can be implemented with a computation model that requires only a finite number of discrete steps (if it ever completes execution). There is no *a priori* bound on the number of steps, but at the time a computation completes, only a finite number of steps have been performed.

This leads us to an important result concerning feedback and feedforward computing networks.

Definition: A *Feedforward Computing Network* implements a finitely implementable function $Z = G(X)$. $G(X)$ has a finite number of inputs $X = < x_1..x_m >$ and outputs $Z = < z_1..z_p >$. The implementation consists of a finite non-recurrent composition of well-defined partial functions $F_1..F_n$. Each of the individual functions Fi maps finite input vectors to a single output. All values, whether primary inputs, outputs, or intermediate are members of an enumerable set. When shown graphically (Figure 2.1), the implementation composes the functions F_i in a feedforward manner. Information flows from the primary inputs (X), through the functional blocks, to the primary outputs (Z).

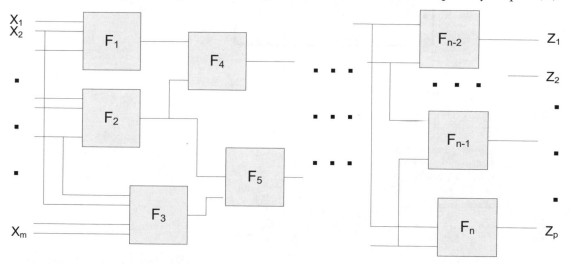

Figure 2.1: A feedforward network can be specified as a finite set of functional blocks, each of which implements a computable function.

We are very familiar with implementations of this type. Combinational logic networks are a good example; the functions F_i are implemented as logic gates. Implementations of feedforward artificial neural networks also belong to this class; the F_i are typically sigmoid neurons. Also included are feedforward dataflow computers, where the F_i are arithmetic and logical operations.

Although it might appear that restricting such networks to feedforward flow reduces computational power with respect to networks containing feedback (i.e., recurrent networks), this is not the case.

In a system containing feedback (Figure 2.2a), the output is a function of a finite, but unbounded sequence of input vectors, $X_1, X_2, \ldots X_s$. The (uppercase) X_i indicate vectors. The sequence of input vectors must be finite to satisfy Turing finite time constraints, i.e., any Turing computation finishes after a finite (but unbounded) time, if it completes at all. Because the sequence of vectors is of finite length, s, the feedback network can be unrolled s times to yield an equivalent feedforward network (Figure 2.2b).

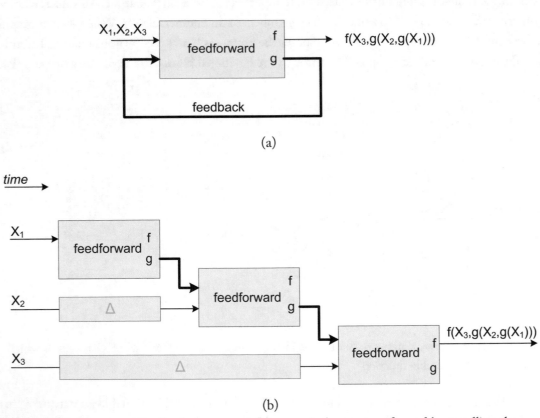

Figure 2.2: (a) A recurrent feedback system and (b) an equivalent system formed by unrolling the feedback system.

In short, unrolling is performed by identifying and cutting a complete set of feedback signals (produced collectively by function g in the figure). Then, replicas of the original network are created and connected in a feedforward fashion. In the drawing, heavy lines indicate feedback signals in the original network. Consequently, although the feedforward network is ostensibly "feedback free," in fact functionally equivalent signals are still present.

Because we are implementing only computable functions, as discussed at the beginning of the subsection, any such recurrent network must complete after a finite number of steps, s (if it completes at all). After unrolling, the feedforward network consumes finite space and requires finite time for the given computation. Because the unrolls are replicas, the specification is *a priori* finite and is the same regardless of the unroll factor. The unrolled feedforward network therefore implements a Turing computable function that is equivalent to the one implemented by the feedback network.

Of course, implementing such an unrolled network might be extremely inefficient with respect to the equivalent rolled, recurrent version. That is, the unroll factor for a given function and associated input, although finite, may be extremely high. Efficiency isn't at issue here, however, only computing capability. The critical point is: *limiting consideration to feedforward networks loses no computational generality with respect to feedback networks.*

This fact will allow us to study feedforward computing networks as a way of understanding the basic computational paradigm(s) as used in the neocortex, without being concerned about losing any computational generality. After developing a basic paradigm, one can later strive for greater efficiencies by using feedback networks based on the same basic computing paradigm.

2.3   GENERAL TNN MODEL

Definition: A TNN is a feedforward connection of neurons, where each neuron implements a well-defined computable function having a finite number of inputs and a single output. Collectively, the feedforward connected neurons evaluate a function with m primary inputs X = < $x_1..x_m$ > and p primary outputs Z = < $z_1..z_p$ >. The input, output, and internal network values are positive integers (or scaled fixed precision numbers). These values are interpreted as spike times. Informally:

1. The neurons are self-timed: the incoming spikes trigger the generation of an output spike. Therefore, a neuron's output spike time can occur no earlier than its first input spike time. Furthermore, a spike is never generated spontaneously; there must be at least one input spike that precedes it.

2. Neuron operation is time-invariant: if the input spikes are all time-shifted by some constant c, then the only change to the output is that the output spike is also time shifted by c.

2.4   SPACE-TIME COMPUTING SYSTEMS

TNNs employ spikes and spike time as part of their description. However, from a mathematical perspective, the temporal metaphor is not needed. In fact, we can define a broad class of feedforward computing networks that are isomorphic to TNNs.

A **Space-time Computing System** is a feedforward computing network composed of a finite set of computable functions F_i defined over set S, where S consists of 0, the set of natural numbers, and ∞. S is closed under addition. For any $x \in S$ $x + \infty = \infty$. $x < \infty$ for any $x \in S$, other than ∞.

$$F_i: (x_1..x_q) \to z; \qquad x_{1..q}, z \in S$$

All functions F_i satisfy:

causality: If $x_j > z$, then z is not affected by x_j

and $z \geq \min(x_1..x_q)$

invariance: $F_i(x_1 + c, .. x_q + c) = F_i(x_1,..x_q) + c$ for any constant $c \in S$.

The F_i functions map a vector of inputs $x_1..x_q$ to an output z. Inputs and outputs are non-negative integers or ∞. Although defined for non-negative integers, one can use simple scaling to allow implementations based on any rational non-negative numbers for inputs and outputs. There are no constraints on the way the F_i are specified and computed; internally, the F_i could be specified via Matlab code, for example.

Causality says that an output value z is affected only by inputs that have a lesser or equal value. Because z is not affected by input $x_j, > z$, then x_j can be set to any value greater than z, and the value of z does not change. Finally, for causality to hold, z must greater than or equal to at least one of the input values.

Invariance says that if all the inputs of F_i are displaced by the same constant c, then the only change in the output is that it is also displaced by c.

* * *

Note that in TNNs as defined here, there are few restrictions on the function that a "neuron" performs. The function does not have to model a biological neuron in any way, although typical models for biological neurons do satisfy the restrictions. A modeler can choose any computable function to be implemented by a given neuron, as long as its primary input/output behavior satisfies the space-time constraints.

The isomorphism between TNNs and the space-time computing networks should be apparent. This is illustrated in Figure 2.3. When one interprets the members of set S to the spike times of a TNN, the TNN operates in a completely natural and intuitive way, with spike times behaving in accordance with the passage of time. For example, causality captures the self-timed nature of neuron operation.

Although we are studying space-time paradigms as a general proposition, in the remainder of this book the TNN interpretation and terminology will be used without losing any generality. For example, with the TNN interpretation of space-time computing, the "invariance" condition becomes "temporal invariance".

A fundamental assertion of this book is that evolution used time as a communication and computation resource, and this led to time-dependent implementations for cognitive functions in the neocortex. It is also asserted that these time-dependent implementations readily support cognitive capabilities that are otherwise extremely difficult or impossible to conceptualize with conventional time-independent implementations. This book is essentially an early step in the exploration of space-time computing models (in the form of TNNs) and their cognitive capabilities.

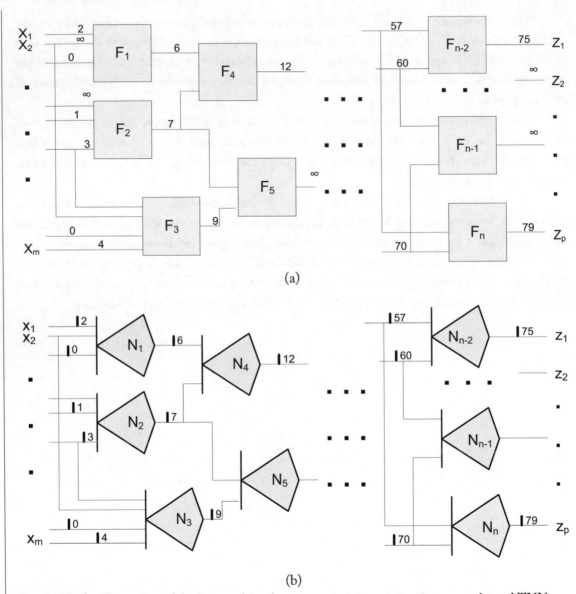

Figure 2.3: An illustration of the isomorphism between space-time computing networks and TNNs. (a) A space-time network with values assigned to inputs and outputs. These values satisfy all the space-time network constraints. (b) An isomorphic TNN. All the space-time values become spike times. An ∞ value becomes a non-spike.

2.5   IMPLICATIONS OF INVARIANCE

Invariance is an important property that deserves further discussion. In TNN terms, each neuron only starts computing with its first input spike arrives, and the computation is completely local with all input times being defined relative to the time of the first spike. As will be seen later, this local computation property carries over to training as well, i.e., training is also local and is based on invariance.

Locality of operation resulting from invariance has interesting implications on implementations where the passage of time is simulated. Consider the function, N_i, performed by a single generic neuron. Because of time invariance, we can decompose the implementation of N_i into two parts. One part, the *global time wrapper* converts global time inputs to local time inputs, and conversely converts a local time output to a global time. The other part implements the local neuron function N_i'. That is, for an arbitrary input vector $< x_1..x_m >$, let $c = x_{min}$. Then we only need to implement $N_i'(x_1 - c,.. x_m - c)$. This guarantees that at least one of the inputs (the one that is originally x_{min}), will be a 0. So, we need implement N'_i for only a restricted set of input times. Then, to evaluate N_i for an arbitrary input vector, the global time wrapper finds the minimum input time and subtracts it from all the input times, the new, normalized times are applied to the implemented N_i', and the global time wrapper adds the minimum time back into the result. A simple example is in Figure 2.4.

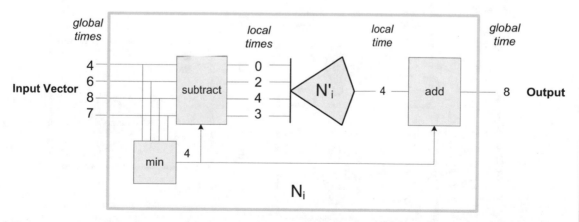

Figure 2.4: Example of N_i implementation which takes advantage time invariance. The implementation consists of a localized neuron implementation N_i' and a global time wrapper (everything else).

Next, consider a simple network made of three neurons (Figure 2.5). The neurons implementing N_i is the same as in Figure 2.4.

In Figure 2.5a, a set of primary inputs are applied, the input spike times, expressed as a vector are shown at the left. The two neurons N_1 and N_2 convert their inputs to a local time vector (as in Figure 2.4). After computation, the local times are converted back to global time and are passed to N_3.

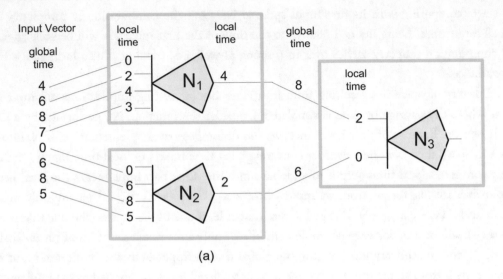

(a)

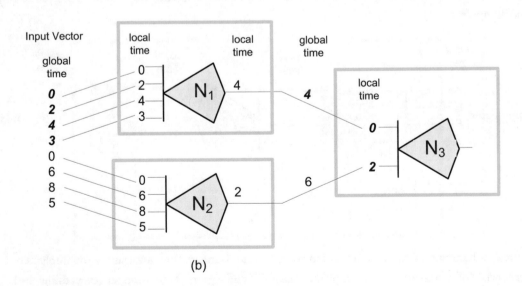

(b)

Figure 2.5: Three neuron networks. In networks (a) and (b) the global time inputs to N_1 change, but the local time inputs do not. Changed times are shown in bold italics. The global time inputs to N_2 do not change. The change in N_1 inputs alone cause both local time inputs to N_3 to change.

In Figure 2.5b, the primary input vector is changed with respect to Figure 2.5a; the elements that feed N_1 are uniformly displaced by 4. Meanwhile, the elements that feed N_2 are unchanged. Within N_1's local timeframe, however, its inputs are unchanged, so its local time output is the same. However, its global time output does change, reflecting the change in its global time inputs. Meanwhile, the inputs to N_2 are identical to their times in Figure 2.5a, so its outputs are identical.

Next, observe the inputs to N_3. The local inputs to N_3 are changed—but not only the inputs that come from N_1, but also the inputs coming from N_2. In other words, changing times through the N_1 path also changes the observed inputs coming from the N_2 path. These changes reflect an automatic adjustment of relative times through the maintenance of global time.

In effect, the minimum time subtracted at the inputs and then re-added at the outputs is global information that is modified by some local increment as it passes through a neuron. Each functional block computes within its own local time frame (the first spike determines t = 0), and the global time frame places all the local information into a common context. In the example, the effect of global time is evident in the changes to both locally observed inputs of N_3, not just the input downstream from the changed inputs.

2.6   TNN SYSTEM ARCHITECTURE

Without loss of computational generality, we consider only feedforward TNNs. However, a TNN alone is not sufficient for constructing a useful computer system. A TNN as envisaged here is a compute engine that operates entirely on temporally encoded information. To construct a useful computing system, we may need to deal with system inputs and/or outputs that are not in temporal (spike) form. This leads to an overall system architecture with information encoding and decoding at the primary interfaces of a feedforward TNN.

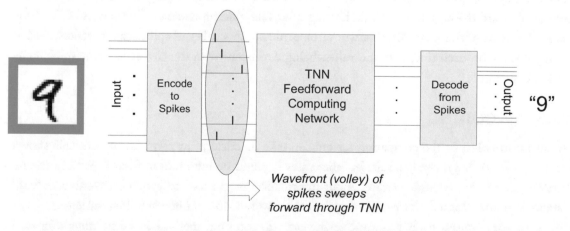

Figure 2.6: Feedforward TNN-based computing system. Conceptually, computation is performed as a single wavefront, or volley, of spikes sweeps through the network.

Figure 2.6 illustrates the full implementation of a cognitive function—the one that is the subject of an extended design study in Part III of this book. The example function is recognizing handwritten Arabic numerals (0–9). In this example, the input is a grayscale image of a handwritten numeral and the output is an indication of which numeral it is, in some easily human-accessible format. It could be given as a 1-out-of-10 binary code for example.

2.6.1   TRAINING

The functions implemented by the individual functional blocks (e.g., neurons) comprising a TNN are determined via a training process. Training will be described in later chapters, but in very general terms, training consists of applying input patterns so that internal TNN structure may be modified to reflect features of the patterns. Training most obviously affects synaptic weights, but there are other neuron and network features (including network interconnection structure) that may be adjusted as part of an overall training procedure.

With *supervised training*, the network is given both input patterns and some additional meta-information regarding the patterns. In the example of Figure 2.6, meta-information may be a label indicating the Arabic numeral that is in the input pattern. With *unsupervised training*, only the input patterns are available when training is performed, there are no labels or other input meta-information. In this book, we strive for implementations that use unsupervised training, although there are some situations where meta-information is used. Decoding, to be described in Section 2.6.4, is one of them.

2.6.2   COMPUTATION (EVALUATION)

The computation process performed by a TNN consists of a single wave (or volley) of spikes that sweeps forward through the network. During this evaluation process each line carries at most one spike. In terms of the space-time network isomorphism, this volley of spikes can be represented as a single value on each line, with the values being consistent with the two space-time implementation constraints.

2.6.3   ENCODING

Input information to the computing system can take practically any form. In the example shown in Figure 2.6, it is grayscale pixels. In other cases it might be information from a grid of pressure sensors on an aerodynamic surface. Input information might also be symbolic, words like "cat," "stalks," "mouse," for a TNN performing the cognitive function of understanding sentences.

In some cases, such as grayscale images, an encoding method is fairly straightforward (although perhaps not as straightforward as one might first think—Section 9.3). In other cases, coming up with an encoding method may be a very difficult part of the overall design process—how would one encode English words as temporal spikes in a way that is conducive to a TNN computation? In cases such as this, the encoding (and/or decoding) may be 9/10 of the problem, figuratively speaking. It is important to note that in any case the encoding method is tightly coupled with features of the internal TNN functions.

Furthermore, depending on the input information and encoding method, there may be some information loss during the encoding process. Whether information loss occurs or not depends entirely on the specific implementation. As it turns out for the particular example shown, we can encode in the grayscale images in a manner that loses no information, but information loss does occur during decoding.

2.6.4   DECODING

After input encoding and TNN processing, the TNN outputs are a temporal spike pattern. The output encoding effectively captures similarity characteristics of the applied input pattern. If we use unsupervised training, then the TNN implicitly determines the output encoding for any given input. This determination depends on features of the input patterns, the neuron-level training function, the network structure (which may be pseudo-random), and other things. Unfortunately, because we as the system designer don't have a way to directly specify an output encoding method *a priori*, the implicitly determined, temporally encoded output method may be (and most likely *will be*) completely incomprehensible to us. The output spike pattern will have no obvious connection to a numeral at the input, "9" for example.

In machine learning parlance, output lines define a *feature space*, and a specific spike pattern encodes *features* of the input. With unsupervised training, it is the combination of internal neuron function, network connectivity pattern, and training set that determine the features implicitly, and the meanings of these features may appear undecipherable from the human perspective. It is the job of decoding to take the collection of features and map them to a recognizable and useful form.

The Problem: We have the temporal output coding as determined internally by the TNN, and we want to map the temporal output coding to a more understandable form. Because we don't know the coding method implied by the TNN, we must come up some way of performing the mapping from output spike patterns to useful output. To do this, it seems that relying on meta-data is essential. In effect, the decoding mechanism incorporates some type of supervised classifier. In this work, a simple *ad hoc* decoding method that relies on meta-data will be used. In general, one can use conventional classification methods, as done in [65], for example.

Note that the type of meta-data depends on the application. For example, the meta-data may be as simple labels associated with visual images, or meta-data may be specific actions that should be taken when an input is applied, as would be the case with controlling actuators in a robotics application.

Observe that this decoding problem only happens at an interface with the external world. For example, if we present a grayscale image of a handwritten 9, we would like the output to clearly indicate "9." In contrast, if the outputs of a TNN are the inputs to other downstream TNNs, then they can be directly connected without decoding. The need for decoding goes away.

Direct connection of TNNs occurs when large-scale TNN systems are composed of smaller TNN subsystems. The use of hierarchy and composition is a key part of the model given here. Hierarchy exists within a TNN as we shall see, and it also extends to multiple connected TNNs.

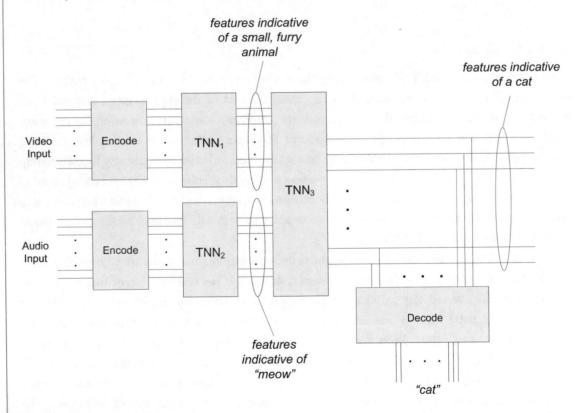

Figure 2.7: TNNs connected without intermediate decode/encode. This is both efficient and suffers no intermediate information loss. This figure also illustrates the binding of features from two sensory inputs (TNN3).

For example, say we have a TNN connected to a video input and another connected to an audio input (Figure 2.7). The video TNN decomposes input images into a feature set, and the audio TNN decomposes sound into a feature set. Then, the outputs of both are fed into a third TNN that analyzes both sets of features. One would *not* ordinarily decode the outputs of the first two TNNs into a humanly understandable form and then re-encode them for use by the third TNN. Not only would this be inefficient, but the decoding/re-encoding would likely lead to needless information loss.

In this example, unsupervised training for the first two TNNs leads to feature sets that may not be humanly decipherable, but the key point is that they are not used directly by a human. Rather, they feed the third TNN, which will be able to use the two feature sets directly.

2.7   SUMMARY: META-ARCHITECTURE

The diagram in Figure 2.8 gives an overall view of the computing structures of interest to us. It contains the three meta-architecture elements defined at the beginning of this chapter. First, as illustrated at the top, we are interested in all the functions that the neocortex implements and will be considering only those with feedforward implementations. However, note that these are equivalent sets, based on finite unrolling of feedback networks to form feedforward networks.

Next, skipping to the bottom of the diagram, we have two types of physical implementations. One type uses the passage of physical time as a communication/computation resource, and the other does not.

In the middle, we have model implementations. Here, there is a division between space-time networks and the non-space-time networks (those computing networks that violate one or both the space-time constraints). The space-time networks (TNNs) maintain the basic temporal features of causality and temporal invariance, consequently they are a natural way of modeling physical implementations that rely on physical time.

In the non-space-time class is virtually every computing method that we currently use. This includes von Neumann computers, for example. All the ANNs from the classical networks to today's deep-learning networks fall into the timing independent category. In contrast, in the space-time network class we have a much smaller set of examples: classical analog computers and TNNs.

If, as is assumed in this work, the neocortex relies on a time-dependent physical implementation, then it is the path shown with heavy arrows that we are most interested in. That is, there is a path connecting function to a space-time model to a time dependent physical implementation.

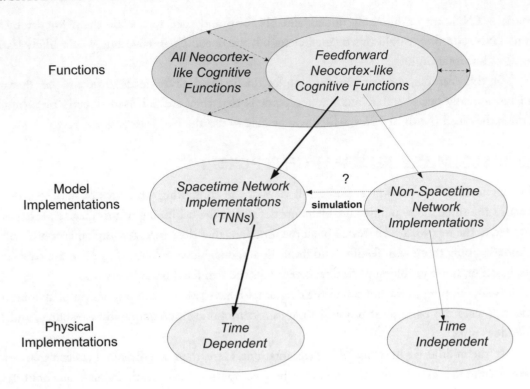

Figure 2.8: Meta-architecture framework.

2.7.1   SIMULATION

Because a function can have multiple implementations, it is possible that any of the cognitive functions also has a non-space-time implementation model. In fact, Figure 2.4 illustrates a non-space-time implementation for a space-time neuron. When we simulate a space-time model, we surround all the elements with a global time wrapper, thereby creating a non-space-time model, which we can then implement on a time independent physical platform (a desktop PC running Matlab, for example).

Given the non-space-time neuron equivalent in Figure 2.4 observe that there are two places where computation takes place, one is in the localized neuron functional block N_i'. The other is in the global time wrapper surrounding the neuron that maintains global time by finding the minimum of the inputs, subtracting the minimum from all the inputs, and adding the minimum back into the output.

Next, consider a time dependent physical implementation, such as the biological neocortex, where the same global time wrapper and N_i' functions are implemented conceptually, but not literally. In this implementation, maintaining global physical time happens naturally. So, in effect, the computations of the global time wrapper functions are completely free. Given that the global time wrapper function is performed at every neuron, and a neuron may have a large number of inputs, this is not an insignificant amount of computation.

2.7.2   IMPLIED FUNCTIONS

When we use a digital computer in the conventional way, we typically start with a function we would like to implement. Sorting a list of integers is a simple example. We then develop an algorithm (or consult Knuth vol. 3), and set about converting it to an implementation. This is typically done by working top-down, writing procedures in a high-level language. The high-level language program eventually compiles down to machine code and is executed on a processor that is also designed in a top-down manner, beginning with an architecture specification.

However, when dealing with cognitive functions, we typically don't have a well-defined function specification at the top. For example, we may want to implement a method for recognizing faces. Face recognition is a well-understood description of a task, but it is far from a well-defined function. How would one go about constructing a clear, unambiguous mathematical definition of "face"?

Despite the absence of a well-defined function, researchers in machine learning have been successful with problems like face recognition. For example, in deep networks, they basically construct layers of sigmoid neurons combined with pooling functions, train them, and test them to see how well they perform. If they perform up to a certain level based on test results, then they are considered to be face recognizers.

A deep neural network used for face recognition does perform a function. If an input is applied to the network, it produces an output. However, in this case, the function is implied by the implementation. Any given model implementation has an *implied function* that is defined by that implementation.

In constructing neural networks (of almost any type), it often happens that the only precise function specification available is the one implied by the implementation. For example, if one were to ask a machine learning developer to describe the precise function of a deep neural network, the answer would be something along the lines of "its function is that thing which it implements." There simply is no higher level, well-defined description of the function. The only precise description is the implementation.

The same is the case with TNNs studied in this book, so implied functions will appear throughout. There are several instances where we don't have a good definition for a function other than its implementation. The implementation typically evolves through an informed trial-and-error process, based on training/evaluation cycles. In these cases, we can state the general task that we would like to perform *a priori*, but not the exact function. Only through a process of building and testing against benchmarks do we arrive an implementation that we deem satisfactory. And, only then, do we have a definition of the implied function being performed.

> *In conventional top-down computer design, implementation follows function. Here, it is function that follows implementation.*

2.8   SPECIAL CASE: FEEDFORWARD MCCULLOCH-PITTS NETWORKS

Consider a TNN composed of simple neurons that employ special-case input and output sets: $S = \{0, \infty\}$. That is, at a primary input or output there is either a spike at time = 0, or there is no spike (∞).

Internally, the neuron's operation is defined by a set of weights, one for each input, and a threshold. Because they are internal and are part of the neuron's definition, the weights and threshold can range over a potentially wide set of values. In the neuron's functional definition, an input spike is mapped to a value of 1; the absence of a spike is mapped to a value of 0. The neuron implementation then forms the inner product of the inputs with their corresponding weights, and if the result is greater than or equal to the threshold, then the output is a spike at time = 0. Otherwise the output is ∞.

Some readers will readily recognize that after converting from spike times to 0's and 1's, this neuron is essentially a McCulloch-Pitts neuron [67], one of the first artificial neurons to be proposed.[2]

We can construct a TNN of these simple neurons (Figure 2.9). These TNNs fit the space-time rules as a special end-case. In these special case TNNs, it is as if all the spikes happen simultaneously at $t = 0$. Note that although this special case satisfies space-time constraints of causality and invariance, it does not really use time as a resource because everything happens at time = 0.

Esser et al. [25] implement networks of this type on the IBM TrueNorth platform [1]. That implementation uses neurons as just described to construct a convolutional neural network. Training is done offline on a conventional computer system using backpropagation methods and more complex neurons, with the final computed weights being transferred to the TrueNorth platform for evaluation.

---

[2] The original source is the one cited, but here a more modern explanation is given.

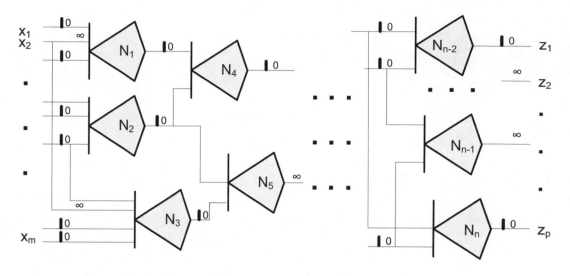

Figure 2.9: Special case TNN where inputs and outputs $\in \{0, \infty\}$.

2.9   RACE LOGIC

Race Logic (RL) is a computing method implemented with conventional CMOS, that employs the precise timing of logic level transitions for encoding and processing information [56, 57, 58]. A feedforward RL implementation is composed of binary logic gates, and an essential part of the paradigm is the effect of inter-gate delays on computed results. Consequently, specifying and controlling these delays is a way of determining the function implemented by a RL network.

RL, in a general form, is an isomorph to TNNs; the main difference is that one uses the times of $0 \to 1$ and $1 \to 0$ logic transitions to convey information while the other uses spike times. Hence, RL is another example of a *space-time computing network*. Because RL and TNNs have been developed independently, each offers valuable perspective to the other. In particular, the isomorphism suggests some very interesting ways that TNNs and RL may eventually be co-developed. TNNs provide a framework for developing brain-like cognitive functions, and RL may provide an implementation platform that uses CMOS digital circuits. This approach is not pursued further in this book, but RL offers a very promising method for the direct implementation of cognitive functions.

CHAPTER 3

Biological Overview

At this point, we have a relatively simple model, TNNs, and the claim is that these TNNs capture the basic cognitive paradigms used in the neocortex. But, is it possible that a model for the neocortex can be this simple? If we compare the simple TNN with the incredibly complicated biological neocortex, with its wide variety of neuron and synapse types, with its complex feedback-laden interconnection structure, with all the physical variation in dendrite and axon structures, there seems to be an extremely wide chasm between the model and reality. At this point, any skepticism on the reader's part seems justified.

The matter of closing the gap between reality and the model is the subject of this chapter and the next, which should be considered a pair. The focus of this chapter is on results from neuroscience that support the TNN computation model used in this book. These are selected results that are not intended to be a comprehensive summary of what is known about the mammalian brain. Chapter 4 will draw on results and observations from this chapter to establish the TNN model and features of an overall framework for studying them.

3.1 OVERALL BRAIN STRUCTURE (VERY BRIEF)

Generally, the brain is divided into three major parts. The *brain stem* connects the spinal cord to the rest of the brain. It is the most primitive part of the brain and controls basic functions like breathing and sleep.

The *cerebellum* is located at the base and the back of the brain (see Figure 3.1). The cerebellum is very important for basic motor control, maintaining balance, for example. It also appears to play some role in attention and language (at least in humans). The brains of all vertebrates have a cerebellum.

The *cerebrum* is the part of the brain that is of interest to us. Most of what is pictured in Figure 3.1 are parts of the cerebrum. The cerebrum includes the cerebral cortex, or *neocortex*. All mammalian brains have a neocortex, and the neocortex is largely responsible for a wide range of cognitive functions. These include sensory processing and perception, reasoning, conscious thought, language, and motor commands.

The human neocortex is a thin sheet of neurons that covers the outer surface of the brain. The thin sheet contains folds and ridges, which result in an increased surface area. The surface area is about 2,500 cm^2 (about 375 in^2). It is 2 to 4 mm thick and contains about 100 billion total neurons, or about 40 million per cm^2.

The cerebrum encloses a number of sub-cortical components, two of which deserve at least brief mention (not shown in Figure 3.1; they are hidden by the cerebrum). The *hippocampus* plays a role in forming memories. The *thalamus* is located between the neocortex and the cerebellum. Among other things it relays signals from the senses (except olfactory) to the neocortex. That is, it serves as a hub for sensory communication.

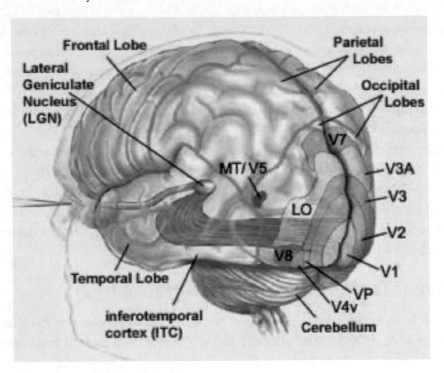

Figure 3.1: The brain, illustrating components belonging to the vision system. From www.thebrain. mcgill.ca. Used with permission.

3.2 NEURONS

An overview of biological neurons was given in the introduction (Section 1.1). As stated in that section, there are two basic types of neurons in the neocortex: *excitatory* and *inhibitory*, of which about 80% are excitatory. Both types are pictured in Figure 3.2. Most of the excitatory neurons are pyramidal neurons, exemplified by the large neuron in the center of Figure 3.2.

Generally speaking, and as used in this book, excitatory neurons perform active computation. Excitatory neurons can maintain relatively precise output spike timing. Their axons can reach both near and far.

In contrast, inhibitory neurons, often referred to as "interneurons,"[3] perform what one might consider shaping or filtering functions on excitatory spike trains. Their output spike timing is less precise than excitatory neurons, and they tend to interoperate with virtually all other neurons in the same local volume. Groups of inhibitory interneurons consequently appear to function collectively, to apply a "blanket" of inhibition [45].

Note that inhibitory neurons are also referred to as "GABA-ergic" for the synaptic neurotransmitter that is involved in their operation. A neurotransmitter is the chemical substance that bridges the synaptic gap connecting an axon with a dendrite. So, the terms "inhibitory neurons," "interneurons," and "GABA-ergic neurons" are often used synonymously, although none is exactly equivalent to either of the other two.

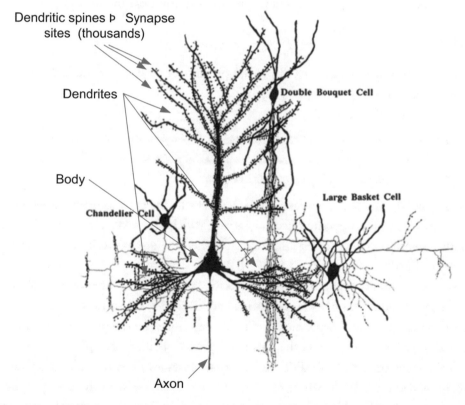

Figure 3.2: An excitatory pyramidal neuron (lower, slightly left of center, unlabeled) surrounded by three inhibitory interneurons. Drawing reprinted from DeFelipe and Fariñas [22]; notations for pyramidal neuron components added by the author (blue arrows). Copyright © 1992, used with permission from Elsevier.

---

[3] Although a small minority of interneurons are excitatory.

3.2.1   SYNAPSES

Synapses are more than mere "connection points" as a computer engineer might think of them. As will become apparent, synapses play a critical role in computation. Synapses may be electrical or chemical, with electrical acting faster than chemical. The description of synapse operation in Section 1.1 assumes chemical synapses. Unless otherwise noted, the synapses in this book are assumed to be chemical.

Individual synapses are either excitatory or inhibitory with respect to the neuron they feed. Furthermore, *Dale's law* (or Dale's principle) states that the synapses driven by a given upstream neuron are either all excitatory or all inhibitory with respect to all the downstream neurons; they are not mixed. Consequently, one can refer to an "excitatory neuron" (all the synapses it drives are excitatory) or to an "inhibitory neuron" (all the synapses it drives are inhibitory).

3.2.2   SYNAPTIC PLASTICITY

Although a number of neural elements display adaptability or *plasticity* of one form or another, the most commonly studied is synaptic plasticity [90].

Donald Hebb [37] observed:

When an axon of cell A is near enough to excite cell B or repeatedly or consistently takes part in firing it, some growth or metabolic change takes place in one or both cells such that A's efficiency, as one of the cells firing B, is increased.

This a general, qualitative statement, which speaks to the enhancement of synaptic efficacy or *potentiation*, but not the opposite effect, *depression*, which also takes place. In 1973, Stent [95] studied both potentiation and depression, and established the groundwork for *spike timing dependent plasticity* (STDP).

Experimental results supporting synaptic plasticity for both potentiation and depression appeared in the late 1990s, from Markram et al. [62] and Bi and Poo [3]. These experiments showed that synaptic potentiation and depression are highly dependent on the precise time difference between a spike reaching a synapse and the downstream neuron emitting a spike.

The classic illustration of STDP for excitatory synapses is in Figure 3.3. In this figure, neuron j is upstream of neuron i. The x-axis is the time difference, Δt, between the pre-synaptic spike t_j^f and the post synaptic spike t_i^f. If Δt is negative, then the spike from the upstream neuron precedes the downstream neuron's spike. In this case, the synaptic efficacy, or weight, is increased as shown on the y-axis. The change in amplitude of later excitatory responses is therefore increased. If Δt is positive, then the upstream neuron's spike occurs before the downstream neuron's spike, and the synaptic weight is depressed. Furthermore, the greater the absolute value of Δt, the smaller the change. In the region where Δt transitions from negative to positive, there is a relatively sharp shift from potentiation to depression.

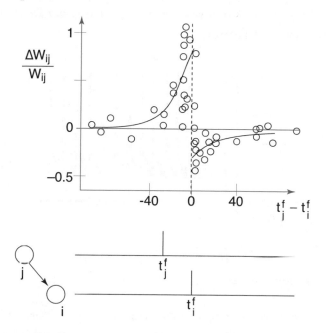

Figure 3.3: Experimental data supporting STDP. Data from Bi and Poo[3], drawing from Gerstner and Kistler [30], copyright © 2002 Cambridge University Press.

The commonly used intuitive argument explaining this behavior is that if the upstream neuron's spike precedes the spiking of the downstream neuron, then it contributes to the occurrence of the downstream neuron's spike and its influence should be enhanced in the future (by increasing the synaptic weight). On the other hand, if the upstream neuron's spike occurs after the downstream neuron's spike, it had nothing to do with the downstream neuron's spike and its influence should be diminished in the future (by decreasing its synaptic weight).

3.2.3   FREQUENCY-CURRENT RELATIONSHIP

In normal operation, a biological neuron receives trains of spikes through its inputs (synapses to dendrites), and if enough spikes are close together in time, the neuron will generate an output spike at its axon. If there is a high level of input stimulus, e.g., lots of inputs spikes very close together in time, the neuron will spew spikes. The greater the stimulus, the higher the output spike rate, eventually limited by the refractory time.

This relationship was recognized very early on, and became the basis for characterizing individual neuron behavior. Because spikes at synaptic inputs lead to current flow into the neuron body, the same effect on the neuron body can be achieved by directly injecting current via an experimental electrode. The more the injected current (I), the higher the output spike rate (or frequency (F)). For example, Figure 3.4 shows the output spike train (top) that results from turning on a constant current source (step function at the bottom).

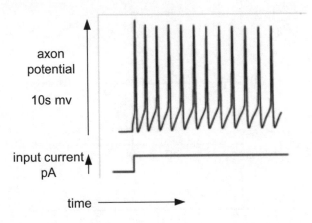

Figure 3.4: Spike train response (top) to current step function stimulus (bottom).

This behavior is summarized in an *FI curve*, where a range of current levels are injected as step functions and the output spike frequencies are measured. A typical FI curve, generated from a model, is in Figure 3.5. This type of behavior appears to support the traditional assumption that the spike rate encodes information: the greater the input stimulation, the higher the output rate. As observed by Laurent [48], "Mean rate interpretations, however, are often simply a consequence of experimental conditions in which a constant stimulus is sustained for a long time."

Furthermore, using any of the basic neuron models, it is also the case that the greater the stimulus (the applied current), the sooner the first spike appears in response. This suggests that the time of the first spike alone may be capable of encoding information.

So: (1) the greater the stimulus, the higher the frequency, and (2) the greater the stimulus, the sooner the first spike. These two strongly correlated relationships have led to two apparently different assumptions regarding the encoding of information in the neocortex: rate coding and temporal coding (Section 1.3). So, although the two encoding methods are quite different, both appear to be consistent with single-neuron FI-based experimental data.

Just because they are consistent does not mean that they are equivalent, however, when it comes to communicating and processing information. Experimental results that draw a clear distinction between rate-based and first spike-based coding are described and discussed in some detail in Section 3.5.

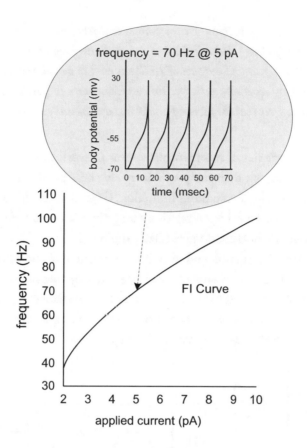

Figure 3.5: Formation of FI curve. Spike response to input current of 5 pA (top) contributes a data point to FI curve (bottom).

3.2.4   INHIBITION

Inhibition has the opposite effect of excitation on the membrane potential. That is, the response to an input spike through an inhibitory synapse causes a neuron's membrane potential to decrease.

There is a wide variety of inhibitory neurons, many with descriptive names (three examples are noted in Figure 3.2), and they are thought to perform a variety of functions. For example, one of the inhibitory functions is to lend stability by balancing excitation, thereby preventing runaway spiking, linked with epilepsy in humans.

As noted previously, inhibitory neurons tend to make dense connections with neighboring excitatory neurons. These connections with neighboring neurons appear to be non-specific, and the neurons appear to operate *en masse*. Of the wide variety of interneurons, two types of greater interest to us here are fast spiking (FS) and low threshold spiking (LTS) varieties [4]:

Our results show a striking dichotomy: the synapses that form both the inputs and outputs of FS cells are initially strong and reliable but depress during repetitive activation; in contrast, the synaptic inputs and outputs of LTS cells tend to be quite weak and unreliable until facilitated during repetitive activity. The complementary properties of these two systems of inhibition suggest that they serve parallel but dynamically distinct roles in neocortical function.

That is, these two varieties of interneurons appear to work cooperatively: the FS cells turn on very quickly and strongly, but then their inhibitory effect fades. Meanwhile, the LTS cells turn on more slowly but are capable of sustaining inhibition. Overall, then, inhibition can be turned on very quickly, and it can be sustained for a relatively long time if needed. Furthermore, virtually all excitatory neurons in the neighborhood are similarly inhibited.

The inter-operation of excitation and inhibition is essential to the TNN paradigm described in this book. One can isolate basic forms of inter-operation by focusing on *disynaptic* micro-circuits consisting of an inhibitory interneuron and one or two excitatory neurons (Figure 3.6). These micro-circuits contain two synapses each, hence the name. The figure contains three different disynaptic micro-circuits, overlapped in the same drawing.

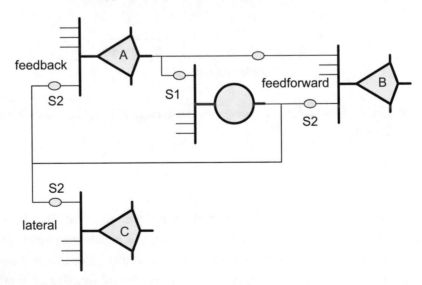

Figure 3.6: Three types of disynaptic inhibition. For all three types S1 is the first synapse, and each of the three types has its own synapse S2.

In all three circuits, the first synapse (labeled "S1" in the Figure 3.6 schematic) is between excitatory neuron A and the inhibitory neuron. A second synapse (all labeled "S2") is between the inhibitory neuron and an excitatory neuron. Three possibilities for S2 are illustrated simultaneously in the figure. With *feedback* inhibition, the disynaptic path is from neuron A, through the inhibitory neuron, back to A. With *feedforward* inhibition, the disynaptic path is from neuron A, through the inhibitory neuron, to neuron B, which also has a direct connection with neuron A. With *lateral* inhibition, the disynaptic path is from neuron A, through the inhibitory neuron, to neuron C, which does not have a direct connection with neuron A.

Although abstract at this point, more intuitive, higher-level functions implementable via these primitives will be given later (Section 7.4.2).

3.3   HIERARCHY AND COLUMNAR ORGANIZATION

The neocortex has a hierarchical structure. Each level in the hierarchy will be summarized in turn, from the bottom-up.

3.3.1   NEURONS

At the bottom are the individual neurons that communicate via voltage spikes; see Figure 3.7. There are two basic types of neurons in the neocortex: excitatory neurons and inhibitory neurons. The output of one neuron (its axon) is connected to the inputs of other neurons (at their dendrites). Synapses are located at the connection points.

3.3.2   COLUMNS (MICRO-COLUMNS)

At the next two levels of the hierarchy, groups of physically neighboring neurons are organized into roughly cylindrical *micro-columns*, sometimes referred to as mini-columns or simply *columns*. On the order of 100 neurons form a column; see Figure 3.7b. These columns are generally thought to be the basic unit for cognitive computation.

As an example, note the two lists of numbers in small print on the right side of Figure 3.7b, these are the counts of excitatory "non-GABA-ergic" (typically excitatory) and "GABA-ergic" (typically inhibitory) neurons. For a neocortical column in this figure there are 64 excitatory neurons and 16 inhibitory neurons, for a total of 80 neurons.

Within a column, neurons are physically organized into layers that are parallel to the surface of the neocortex. Classically, there are six layers, but this is somewhat arbitrary and depends on the particular region of the neocortex that is being discussed. For example, in Figure 3.7c, layers II and III are combined (and this is often done in illustrating the visual cortex). Meanwhile, some layers are divided into sub layers as illustrated in the figure (layers IVa and IVb). Note that in the literature sometimes Roman numerals denote layers, and sometimes it's Arabic numerals.

In the theoretical neuroscience literature, this six-layer structure receives a lot of attention. In this book, however, the characteristics of the six columnar layers are not used as a starting point for constructing computational circuits. Starting with neural architecture at this level seems analogous to the aliens of Section 1.6 discovering six-layer circuit boards and assuming that six layers of copper wiring are essential to the way conventional computers compute. In other words, physical constraints faced by evolution may have determined the columnar layering structure more than something fundamental to the computational paradigm.

Rather than try to fit a computational architecture into the constraints of biological columnar circuits, in this book a looser, more free-form drawing board is used initially for interconnecting computational neurons. For example, the neurons in a columnar network contain a significant amount of excitatory feedback. However, as argued earlier in Section 2.2, from a computational perspective, there will always be an unrolled feedforward functional equivalent. By freeing ourselves from the constraints of the biological columnar circuits, we hope to simplify the effort by focusing on the simpler feedforward equivalents.

To be clear: It is not being suggested that the columnar structure is not important. It may turn out to be extremely important, especially for higher-level cognitive functions. However, for our purposes, it seems premature to assume that it is essential to the basic computational paradigm without first understanding the basics of neuron-level communication and computation.

Regarding layers, the single feature that is of interest to us here is: important physical pathways through a column often enter at layer IV and exit at V. So, one can conclude that the minimum path length through a column contains at least two serially connected neurons. This little piece of information will be used later in the construction of Figure 3.15 and associated discussion.

3.3.3   MACRO-COLUMNS

At the next level of the hierarchy, on the order of 100 columns are physically grouped into a *macro-column*.

At this hierarchy level, the neuroscience literature sometimes contains ambiguous terms. In particular, the smaller diameter columns are sometimes referred to as "micro-columns" or "mini-columns." The larger columns, composed of multiple smaller columns, are referred to as "macro-columns" or "hyper-columns." Ambiguity occurs because the simple term "column" may be applied to either, depending on the author and the context.

In this book, the simple term "column" is reserved for the smaller, lower level columns and the term "macro-column" is used for the larger columns. Occasionally in this work, "micro-columns" is used as a synonym for "columns," to emphasize the contrast with macro-columns.

Macro-columns also have different names depending on their specific functions; for example, "barrel columns" are often associated specifically with the heavily studied somatosensory cortex of the rat (discussed later in Section 3.5).

3.3.4 REGIONS

Some large number of macro-columns (the number varies widely) are grouped to form *regions*. Figure 3.7d is excerpted from a very well-known schematic drawing of the vision system of a macaque monkey [27]. Some of these regions also are shown in Figure 3.1. Larger regions may be formed of interconnected smaller regions (subregions), as illustrated in Figure 3.7d.

Physically, regions are essentially laid out in two dimensions over the neocortex. The third dimension contains the six columnar layers. The structure is reminiscent of a two dimensional, multi-layer circuit board, except in the neocortex the layers contain active elements (neurons and synapses). Researchers performing fMRI imaging collect data at the region level.

3.3.5 LOBES

The neocortex is composed of four lobes. These are basically the highest-level regions and are denoted in Figure 3.1. Following is a very brief summary.

Frontal Lobe: A number of higher-level cognitive functions take place in the frontal lobe. These include cause-effect relationships such as the prediction of future outcomes from current actions, judgment involving good and bad actions, and assessment of differences and similarities. Personality is formed in the frontal lobe, as is the ability to speak fluently.

Parietal Lobe: The parietal lobe consists of two main regions (left and right). One is responsible for integrating information from multiple senses (touch, temperature, taste) into a single perception. The other deals with visuospatial perception and object manipulation.

Temporal Lobe: The temporal lobe holds visual memories and processes auditory information (language comprehension) and visual information (object recognition).

Occipital Lobe: The occipital lobe is the center for visual processing. Information from the retina flows through the optic nerve, and then to a processing hierarchy of regions in the occipital lobe. This processing pathway is of particular interest in this book and will be discussed in more detail later. This pathway is highlighted in Figure 3.1.

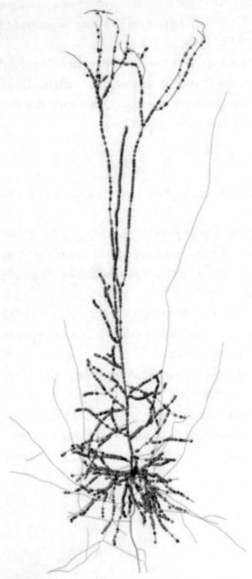

a) Neuron, from Hill et al. [39]. Copyright © 2012 National Academy of Sciences. Used with permission.

(b) Column, from Peters and Sethare [78]. Copyright © 1996 Wiley-Liss, Inc. Used with permission.

Figure 3.7: (a) and (b).

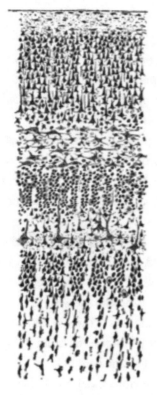

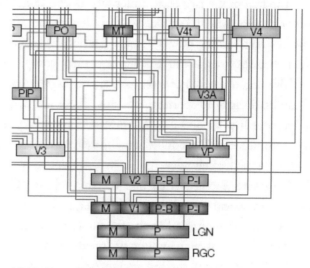

(d) Regions, excerpted from Felleman and Van Essen [27], by permission permission of Oxford University Press.

(c) Macrocolumn, from Ramon y Cajal via Wikipedia

Figure 3.7: Four layers of neocortical hierarchy.

3.3.6 UNIFORMITY

Uniformity provides important leverage for expanding the applicable scope of experimental results. There are two basic uniformity assumptions that underpin the approach described herein: (1) the fundamental cognitive paradigm used by the neocortex is uniform across species, and (2) it is also uniform across brain regions, within the same species.

These assumptions come directly from the *columnar hypothesis* or *columnar theory*, as espoused by Mountcastle [71, 72, 73] and referred to by Hawkins as the "Rosetta Stone of neuroscience"[36]. In 1978 Mountcastle put forward:

> *"...the general proposition that the processing function of neocortical modules is qualitatively similar in all neocortical regions. Put shortly, there is nothing intrinsically motor about the motor cortex, nor sensory about the sensory cortex. Thus elucidation of the mode of*

operation of the local modular circuit anywhere in the neocortex will be of great generaliz-ing significance."

The first of the two basic uniformity assumptions allows us to expand the scope of conclusions drawn from experimental studies of a variety of species. The second assumption is important because it implies that by learning how the simple feedforward cognitive tasks are computationally implemented in the neocortex, we are also learning how more complex cognitive tasks are implemented.

The columnar hypothesis asserts the *uniformity* of neuron computation across all known implementations of neocortical columns. There is a body of researchers that accept the position of uniformity as first espoused by Mountcastle; another group takes a position of non-uniformity, that is, the computational "modules" are different in different regions in the neocortex. But, uniformity is in the eye of the beholder.

To quote Buxhoeveden and Casanova [14]: "Columnar theory postulates a fundamental strategy for cortical organization. However, it does not require identical units." In the research in this book, we are interested in the fundamental strategy, not the specializations that have evolved for specific functions such as sight or touch, nor for the requirements of different species.

In digital design, the interconnection of gates based on Boolean algebra is the fundamental strategy. However, all functional units constructed of logic gates are not identical—far from it. So, the fundamental strategy is uniform, even though the individual functional units are not.

To conclude, DeFelipe et al. [22a] provide this late 19th quotation from Cajal, the father of neuroscience:

"... if the gray cerebral cortex is an aggregation of organs with diverse functions, each of them must possess a special structure, within a [fundamental] plan whose general lines are appropriate for the whole cortex."

It is the fundamental plan, the paradigm, that is of interest here.

3.4   INTER-NEURON CONNECTIONS

In a system where information is carried by voltage spikes, the connections between neurons become computational components. "Connections" includes not only the interconnection topology, but the delays associated with transmitting information via voltage spikes and responses. In a space-time computation regime, these delays are an essential element of information processing.

Spike propagation delays are a function of a large number of features. One obvious feature is physical path length. Other features include axon and dendrite diameters and the number of branch points along a conducting pathway, all of which affect the spike velocity.

Overall, the spike transmission path is very complex, and modeling it in any detail is a daunting task. Fortunately, for the paradigms of interest here, it is not necessary to deal with the details of actual transmission delays, only their net effect is important. For modeling work to come later, it is sufficient to establish that; (1) communication delays between two neurons vary over a fairly wide range; (2) delays are of the same order as neuron computational latencies; and (3) there are multiple paths (with multiple synapses) connecting pairs of neurons. Following paragraphs provide biological support for these three features.

The path connecting two neurons typically consists of an axon, a synapse, and a dendrite (Figure 1.2). Depending on the function, the axon length can vary over several orders of magnitude. Long-distance communication, say in the spinal cord or between major brain regions, is typically performed by axons that are covered with a myelinated sheath which increases propagation velocity by roughly an order of magnitude over un-myelinated axons.

Of more interest to us here is the functioning and communication that takes place within a column or macro-column, rather than long distance communication. Myelin wrapped around an axon consumes additional volume, so densely packed computational neurons in a column tend to have un-myelinated axons. Consequently, in this work we deal exclusively with the slower un-myelinated axons.

Figure 3.8 pictures a small group of pyramidal neurons contained in a column. This picture gives an idea of the distances involved and the potentially large number of interconnections as the mass of axons and dendrites contact each other at a multitude of synapses.

Figure 3.8: Pyramidal neurons in a micro-column. From Perin et al. [77]. Note the 100 μm distance scale in the lower right corner. Copyright © 2011 National Academy of Sciences. Used with permission.

3.4.1   PATH DISTANCES

Hill et al. [39] performed a detailed study regarding connectivity in the rat somatosensory cortex. Of interest here is the measured path distances; see Figure 3.9. Both axon and dendrite path distances vary over a fairly wide range: approximately 100–700 microns for axons and 50–400 microns for dendrites, with a significant number of outliers.

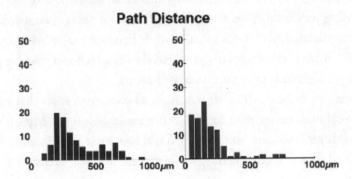

Figure 3.9: Axons are on left in blue; dendrites are on right in red. The x axes are distances, the y axes are percentages. Figure 1e from Hill et al. [39] showing path distances. Copyright © 2012 National Academy of Sciences. Used with permission.

3.4.2   PROPAGATION VELOCITIES

Spike propagation delays are a function of path length and propagation time per unit length. Bakkum et al. [2] electrically stimulated and timed spikes in rat cortical neurons *in vitro* using a multi-electrode array. Data regarding propagation delay are given in Figure 3.10, taken from Figure 1d of [2]. These results indicate propagation velocities of 50–400 microns/millisecond (with a tail). Citing a number of studies, Budd et al. [11] provide an estimate of 100–600 microns/millisecond.

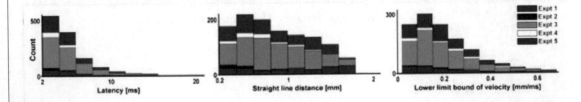

Figure 3.10: Figure 1d from Bakkum et al. [2], showing interneuron latency, distances, and velocities. Copyright © 2008 Bakkum et al. Used with permission.

3.4.3   TRANSMISSION DELAYS

Path lengths vary from near zero to several hundred microns, and velocities vary from one hundred to several hundred microns per millisecond. Furthermore, there is a lot of variation in both velocity and path length. Consequently, when considered in combination, and assuming some degree of independence among the variables, one can plausibly postulate total propagation delays ranging over an order of magnitude: at least from about 1 msec to about 10 msec. Note that latency going through the synapse is remarkably fast, about 1 msec, so most of the transmission delay is in the axons and dendrites.

This biological data supports a key assumption used in this work: *inter-neuron delays are of the same order as neuron time constants (and therefore neuron "computation" time), and the delays vary over a range of at least one order of magnitude.*

The large number of variables and dependence on axon and dendrite structure adds breadth to the distribution of spike propagation delays, and this broad distribution of delays is a key element of the TNN paradigm developed here. At this stage of research, we don't need to know the exact distribution. We only need to know that such a distribution exists.

3.4.4   NUMBERS OF CONNECTIONS

Excitatory neurons in close proximity (as is the case in biological columns) are connected via multiple synapses (Figure 3.11). Markram et al. [63] identified between 4 and 8 potential synaptic contacts between pairs of excitatory neurons. In a more recent paper, Fauth et al. [26] found similar connectivity across a number of cortical layers (Figure 3.12). Note that such studies sometimes undercount because of the experimental methods used; e.g., tissue slicing methods may slice off a portion of the area of interest.

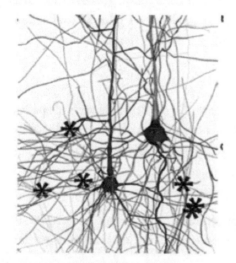

Figure 3.11: An example of multipath connectivity between two excitatory neurons (Figure 3a from Hill et al. [39]). Stars show six connections between the same two neurons. Copyright © 2012 National Academy of Sciences. Used with permission.

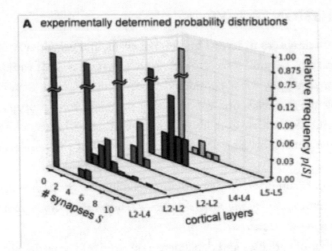

Figure 3.12: Figure 1a from Fauth et al. [26], showing numbers of synapses connecting neuron pairs as a function of the cortical layer. Copyright © 2015 Fauth et al. Used with permission.

3.4.5   ATTENUATION OF EXCITATORY RESPONSES

Excitatory responses are formed in the dendrites, and the dendrites are very complex structures, indeed. The primary feature of interest here is attenuation that takes place as an excitatory response travels from its synaptic source to the neuron body.

Classic cable theory, applied to dendritic propagation, suggests significant attenuation as an excitatory response travels from source to neuron body. However, there are also a number of mitigating factors that greatly reduce the effective attenuation [59, 85]. Rather than attempt to select and model specific propagation/attenuation factors, in the work to follow it is assumed that attenuation may take place, and attenuation effects are collectively modeled via a simple attenuation parameter which is adjusted and tuned as part of the modeling process (Section 6.7).

3.4.6   CONNECTIONS SUMMARY

The bottom line is that if two neurons are connected, then they are (1) connected via multiple paths, (2) each with a different propagation delay, and (3) taken from a fairly broad distribution. This collection of observations is fundamental to the computational paradigm employed herein.

3.5   SENSORY PROCESSING

The research described in this book is based on the assumption that precise spike timing relationships form the basis for communication and computation. Furthermore, the application driver for model development is feedforward information processing. This subsection summarizes experimental results that support the plausibility of both precise spike coding and the existence of feedforward processing of sensory information.

There is more experimental research based on sensory processing than other parts of the neocortex, largely because it easier to perform useful *in vivo* experiments. For example, one can shine simple geometric patterns on a subject's retina and measure the spike responses on the optic nerve (or at other points in the vision system). As another example, the whiskers on a rat's snout are known to map one-to-one with macrocolumns in the rat's somatosensory cortex (Figure 3.13). Because of this association, many experiments consist of tweaking a rat's whisker and observing the spike response via a probe buried in the associated macrocolumn.

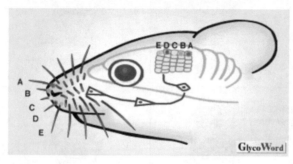

Figure 3.13: Individual whiskers on a rat's snout map 1-to-1 with barrel columns (macro-columns) in the rat's somatosensory cortex. Courtesy of Glycoforum www.glycoforum.gr.jp.

From a research perspective, sensory processing is among the more basic cognitive functions. If we are trying to understand cognitive functions, then it's probably better to start simply. Consequently, as observed earlier, basic sensory processing problems are good drivers for initial research into cognitive computation.

3.5.1 RECEPTIVE FIELDS

Receptors for the senses, broadly speaking, are laid out in two-dimensional space, and a sensory neuron is associated with a small subregion of that space: the *receptive field* (RF) for that neuron. The RF can be a small area of the retina, a patch of skin or tongue. It can be a whisker, or it can be a hair in the cochlea. All the neurons in a column or macrocolumn tend to share the same RF, as is the case for a rat's whisker and barrel columns (Figure 3.13).

The retinal ganglion cells that comprise the RFs in the vision system warrant more discussion. There are two types of cells that take input stimulation from rods and cones: on-center cells and off-center cells. These cells are essentially primitive edge detectors that work as follows (see Figure 3.14).

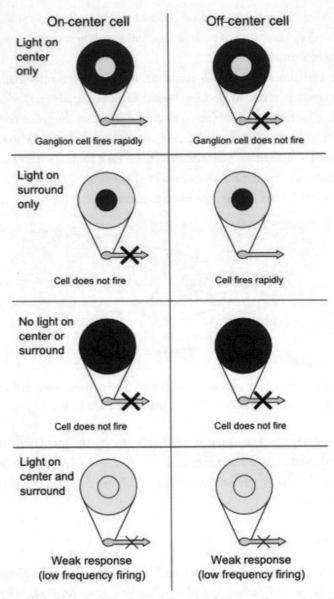

Figure 3.14: On-center and 0ff-center retinal ganglion cell behavior (taken from Wikipedia, https://en.wikipedia.org/wiki/Hypercomplex_cell).

Each of the retinal ganglion cells senses a *center* and a *surround* and drives its axon with a burst of spikes (or not) depending on the light pattern falling on the sensors. If light is more intense on the center of an on-center cell than on the surround, then the cell will emit a spike burst in response. The greater the difference in intensity, the earlier and higher frequency, the spike burst will be. If the surround is more intensely stimulated, however, there will be no spiking (or very little). Conversely, if the surround of an off-center cell is more strongly stimulated, the off-center cell will emit a spike burst, and if the center of an off-center cell is more strongly stimulated there will be no spiking (or very little). In the case of no light, neither will spike. The net effect is that RFs lying along edges of a visual object will yield more spikes from the on- and off-center cells, thereby indicating an edge. Then, neurons in the vision pathway process this edge-detected information.

The above is the conventional explanation of retinal ganglion cells. At this point, however, it is important to recall the strong correlation between spike burst frequency and the times of the first spikes in a burst (Section 3.2.2). One could re-word the preceding paragraph so that references relating spike rates in Figure 3.14 become references to first spike times. For example, in the upper-left corner of the figure, if there is light on the center only, then the first spike in response appears sooner in time.

3.5.2   SACCADES AND WHISKS

Saccades are involuntary or voluntary movements of the eye, which have a number of causes. They are irregularly spaced in time, but may occur at an average rate of a few Hz. Following a saccade, there is a fixation period, during which the on-center and off-center RF cells produce spike bursts corresponding to stimulus intensity, but then fade. Hence, a saccade initiates a wavefront of spike bursts emanating from the receptive fields that are sent to the visual cortex to be processed.

Mammals have whiskers and/or hair as part of their somatosensory system. Rodents in particular have a very well developed somatosensory system (they do a lot of maneuvering in the dark). Whisks in rats aren't just passive—they constantly wiggle their little snouts to turn the whiskers into dynamic sensors, and resulting spike bursts from the whiskers are information rich. As with saccades, active whisks occur at a rate of a few Hz. Furthermore, as noted earlier, individual whiskers are associated with individual macro-columns, which facilitates experimentation.

Beyond saccades and whisks which are singled out in this subsection, there also is evidence of saccadic-like information flow in the olfactory system [100], auditory system [51], and the sense of place [69].

3.5.3   VISION PATHWAY

The pathway in the visual system between the retina (see Figure 3.1) and the inferotemporal cortex (IT) where visual objects are identified has been extensively studied. This pathway is illustrated in Figure 3.15 (derived from Figure 3 in Thorpe and Imbert [97]). Thorpe and Imbert estimate that there are at least 11 neurons (and associated dendritic, synaptic, and axonic delays) between the retina and the first neuron in IT. For example, at least two neurons are needed in V1, V2, and V4 because the neurons receiving the inputs differ from the neurons projecting the outputs because they are in different layers in a column—as noted when discussing micro-columns in Section 3.3.

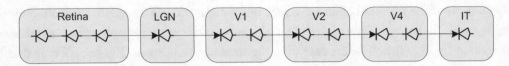

Figure 3.15: Minimum feedforward neural pathway from the retina to the inferotemporal cortex (IT), where visual objects are identified. Signals must pass through at least 11 neurons to make it to the first neuron in the IT.

3.5.4 WAVES OF SPIKES

At the vision processing input interface, saccades appear to divide information flow into waves or volleys. Maldonado et al. [61] monitored the spiking behavior of a small number of neurons contained in the V1 layer of the visual cortex in monkeys. Physically, the V1 layer of a human is shown in Figure 3.1. Figure 3.15 is the shortest feedforward path through the vision system, and V1 is the place in the path where serious computation begins. Hence, any processing that involves vision, no matter how coarse, requires at least some V1 processing.

While viewing natural images, the monkeys performed a saccade typically once every 250–300 msec, with significant variation. After a saccade, there is a *fixation* time, during which the new image is taken in and processed. The red line in Figure 3.16 shows the mean firing rate of V1 neurons during fixation (using a sliding window averaging method) versus time in milliseconds on the x-axis; time = 0 is the onset of fixation.

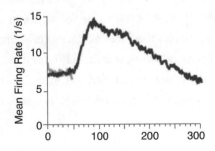

Figure 3.16: Based on Figure 4A from Maldonado et al. [61]. The red line traces the fixation process—the one we are interested in here. Copyright © 2008 by the American Physiological Society.

Following the onset of fixation, the first spikes start to arrive at V1 after about 50 msec (a number consistent with Figure 3.15). Maldonado et al. observed that the rates peaked at about 90 msec and then went back down to the baseline. The drop to the baseline is interesting, because it suggests a sort of reset mechanism. That is, after responding to one saccade, the neurons seem to settle back into a quiescent state, and are ready to respond afresh to the next saccade. This natural drop is largely finished at approximately 200 msec, and the neurons are ready for the next saccade (more or less).

It is worth also noting that at the time a saccade is initiated, a form of inhibition is turned on [84], damping down the excitatory neurons that are active during fixation. The excitatory neurons are thus forced to reset. So, a looping sequence is: (1) trigger saccade and inhibit fixation neurons; (2) perform saccade; (3) turn off inhibition and begin fixation; and (4) fixation neurons accept sensory input and then return to a resting state until the next saccade is initiated (to re-start the sequence).

The above supports the plausibility argument that: saccades (and whisks and other sensory inputs) naturally partition spiking behavior into bursts, albeit in an irregular, aperiodic fashion.

3.5.5   FEEDFORWARD PROCESSING PATH

For decades, Simon Thorpe has been a leading proponent of spike timing relationships for computation and has authored or co-authored a number enlightening papers on the subject (including [33, 34, 96, 98, 103]).

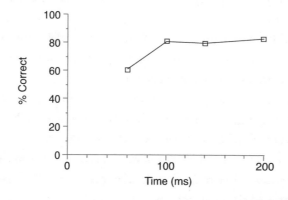

Figure 3.17: Based on Figure 1 from Thorpe and Imbert [97]. The correct classifications of the first of two images presented to a subject. The length of each image presentation is the x axis. Copyright © 1989 Elsevier.

In some of the earlier work, Thorpe and Imbert [97] present experimental evidence that a human subject can categorize (classify) a fairly complex visual image within roughly 100 msec after presentation; see Figure 3.17. In the experiment, a subject was sequentially presented with two images, each for a short time period (measured in tens of milliseconds shown on the x-axis). The subject is asked to identify the object pictured in each image. The graph shows the percentage of correct identifications of the first image, given that the second image is correctly identified. At

60 msec, the image was correctly identified 60% of the time, and at 100 msec, the first image was correctly identified 80% of the time.

Given the experimental data, one can evaluate important timing properties in the visual processing pathway, as illustrated in Figure 3.15. This was done by Thorpe and Imbert, and is repeated here in a slightly different formulation.

The task being performed is one of mapping relatively complex images to a well-defined class in about 100 msec. How can this be achieved with the known visual processing path as in Figure 3.15?

To answer this question, first consider the excitatory neuron computation time. Neurons used as examples in this book are given time constants that are a bit on the leisurely side, and these yield a computational latency of about 10 msec on average (Figure 9.9c). This computational latency is the time between the first input spike in a volley and the time of the first output spike in response. If the time constants are cut in half, the latency through an excitatory neuron becomes a more brisk 5 msec. So, assume 5 msec. If the average path length between two neurons is 500 microns at 200 microns per millisecond (from Figure 3.9 and Figure 3.10), the delay between neurons is 2.5 msec. Finally, the synapse consumes about 1 msec, for a total of 8.5 msec per excitatory neuron on the path.

Then, if there are at least 11 neurons in the path, the total delay through the path is about 94 msec, and the propagation of the very first spikes through the entire path will have barely enough time to meet the time constraint implied by Figure 3.15.

What about the second spikes? The maximum neuron firing rate is about 100 Hz or one per 10 msec (in reality it is usually significantly slower). In any case, it will take at least an additional 10 msec per neuron for the second spike to have any effect, assuming that the first output spike of each neuron depends on both first and second input spikes (the minimum number of spikes to establish a rate). Then the delay per neuron rises to 18.5 msec; a number that is too slow to be plausible, given the experimentally determined time constraints.

In reality, because of the significant jitter in a spike burst, it takes many more than two spikes to determine a rate with any accuracy, so using only two spikes and 18.5 msec per neuron delay for processing rates is very optimistic.

To mitigate the problem with jitter, one might argue that a population of cells driving multiple axons might be used for coding a rate as an average over the population. If so, it takes a very large number of independent neurons (281 to be more precise) to convey a single value with ±10% error [29]. This would be an extremely inefficient coding scheme. If the first spike can carry the information in precise form (see below), why would evolution use 281 neurons to do the same thing? Hence, rate coding—at least for identifying patterns rapidly—seems implausible from an efficiency perspective.

The above argues for using the time of the *first spike* in a burst as the sole information carrier for the burst. That is, from a delay/latency perspective, it appears that there is only enough time for the spike at the leading edge of the burst to pass through the processing path in order to meet an experimentally determined time constraint.

3.5.6   PRECISION

It was just argued that rate coding is implausible for feedforward sensory processing because only the first spikes move through the system quickly enough to be the carriers of information. Being fast enough is not sufficient, however. Also important is whether the relative timing of first spikes is precise enough to carry the necessary information. As it turns out, neurons can maintain relatively high precision for repeated applications of the same stimulus.

Mainen and Sejnowski [60] studied stimulus-response *in vitro* for rat cortical neurons. When stimulated with an input voltage source containing random variability as a way of modeling actual synaptic activity, neuron output spike trains followed a reproducible pattern with high precision between spikes. As summarized by the authors: "stimuli with fluctuations resembling synaptic activity produced spike trains with timing reproducible to less than 1 millisecond." Hence, neurons can support adequate precision for information encoding via single spikes.

In an *in vivo* experiment, Petersen et al. stimulated a rat's whiskers and observed the neural spiking behavior in the corresponding barrel column located in the rat's somatosensory cortex. Figure 3.18, taken from [79], shows the spike times for two different cells in the barrel column for 50 trials (the y-axis). The x-axis is in units of .1 msec, so for most of the trials, the output occurs within a 1 msec time window, thereby providing support for repeatability with 1 msec precision.

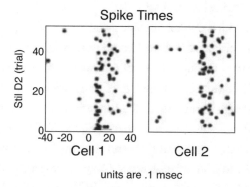

Figure 3.18: Illustrating the repeatability of spike times for two neurons (Cells 1 and 2) belonging to the barrel column whose associated whisker is repeatedly whisked. Each dot in the vertical axis is a trial; the horizontal axis is the relative spike time in response to stimulation. Based on an excerpt from Figure 1a of Petersen et al. [79]. Copyright © 2001 Cell Press.

Finally, Butts et al. [13] show (among other things) that although neurons are capable of 1 msec precision, the time scale of the external visual sensory stimulus is much slower. The ability to temporally discriminate at 1 msec resolution far exceeds the input rate from external stimuli (e.g., via saccadic input). One might reasonably infer that the high neuron resolution is important for internal computation of visual information, not just to support external sensory requirements.

Given that neurons operate at a time scale on the order of 10 msec, and spike times are precise to 1 msec, it appears that there is enough precision to support 3–4 bits of information.

3.5.7   INFORMATION CONTENT

Experimental results show that not only do spikes on individual pathways provide adequate resolution for reliably encoding information, information can be extracted from first spikes generated by a group of neurons responding to the same stimulus.

In an *in vivo* study, Johansson and Birznieks [43] mechanically apply pressure from varying directions to various points on a fingertip and collect neuron spike data from the corresponding receptive fields. As they say in their abstract: "the relative timing of the first-spikes contains reliable information about direction of fingertip force and object shape and provides this information appreciably faster than rate codes."

In an interesting *in vitro* study [46], a culture of rat cortical cells is grown on a substrate with a grid of embedded electrodes. A single substrate has tens to hundreds of thousands of neurons and contains 60 electrodes that can both provide stimulation and record neuron responses (spikes). By random chance, some of the electrodes will make contacts with neurons in a way that is experimentally useful; that is, to be able to stimulate them by injecting current and to measure spike responses.

In the experiment, a single electrical stimulation causes a subgroup of neurons to fire. These are the *immediate* responses which do not involve synaptic activity on their inputs. After the immediate response, activity is propagated to other neurons through synapses; this activity ripples through the network as it passes through series of neurons until it eventually dies out.

In the experiment of interest here, there were five different stimulation points. Beginning 10 msec after the initial stimulus (to eliminate all the immediate responders), spike data was collected for 100 msec, and the resulting spike data was then reduced in four ways. Two of these reduction methods are spike timing based (*time to first spike* and *rank order*) and two are rate-based (*spike count during the 100 msec interval* and a *population count histogram* for a smaller, fixed time window). In Figure 3.19, these are labeled TFS, Rank, Count, and PCH, respectively.

Then, a conventional classifier based on a support vector machine [21] was trained to take the reduced response data as input and predict which of the five stimulation points produced the observed spike response. The accuracy of the classifier is a lower-bound indication of the information content in the analyzed responses. Six experimental trials were performed. Results are plotted in Figure 3.19 (data from [46] supplementary information). Each line indicates the accuracy using one of the four reduced data responses. It is clear that more information is extracted from the precise spike timing data (all have greater than .8 accuracy, some approaching 1.0) than the spike rate data (all less than .8 accuracy). Conclusion: not only do first spike times convey information, they convey significantly more information than rates

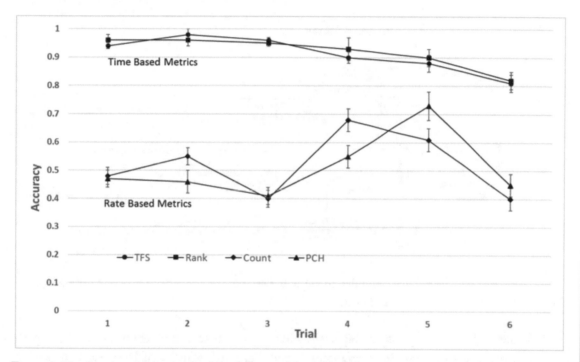

Figure 3.19: showing accuracies for four different data reduction methods; two are time based, and two are rate based. Each line contains results of six experiment repetitions. Based on data taken from Kermany et al. [46]. Copyright © 2010 Society for Neuroscience.

3.5.8   NEURAL PROCESSING

The work by Lee et al. [50] is perhaps the most striking of the experimental results on sensory processing of spike-encoded information.

In that experiment, a trained monkey fixes its gaze on a spot in the middle of a screen. At the time of target onset (see Figure 3.20), a spot[4] of variable contrast appears at a point away from the center. If the monkey's eyes immediately saccade to the new contrast spot, then it is rewarded and a data point is collected. Data collected at the probed neuron in V1 includes the neural latency (NL) and the spike firing rate (FR). Eye movement is also monitored and the saccadic response

---

[4]   A small simple pattern adjusted to stimulate a probed neuron in V1; see [49] for details.

time (RT) is recorded. The neural pathway from target onset to RT is illustrated in Figure 3.20. The portion of the pathway from target onset to V1 where NL is measured is in Figure 3.15. A number of functional blocks in Figure 3.20 are not of direct interest here. The superior colliculus selectively control saccades, especially inhibiting unwanted saccades.

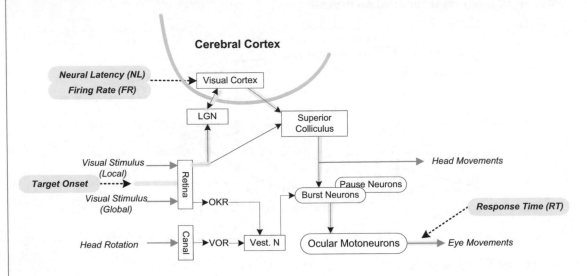

Figure 3.20: Pathway from visual stimulus to saccade. Based on Figure 3 from Hikosaka et al.[38]. Copyright © 2000 The American Physiological Society.

What makes this experiment especially interesting is that it precisely measures a complete *in vivo* pathway from sensory input to motor output. Very little actual visual processing is needed, so the vision part of the pathway may extend no further than V1. The major processing required is the determination of exactly which motor neurons should be activated (and which inhibited) in order to stimulate the muscles to invoke a precise eye movement (the saccade). There are at least five neuron layers from the retina to V1 (Figure 3.15), and several more in the path from V1 through the ocular motoneurons at the lower right of Figure 3.20.

In Figure 3.21, the NL and RT are plotted as a function of the contrast of the target image. For a low contrast target image, the time from the retina to V1 (NL in the graph) is about 85 msec, and for high contrast images it is about 50 msec. These times are consistent with the pathway in Figure 3.15 and the associated time estimates made earlier. Furthermore, the greater the stimulation (the higher the contrast), the sooner a spike arrives at V1.

The RT occurs about 100 msec after NL for a low contrast target image and about 60 msec after NL for high contrast target images. Considering the precise processing required, this strongly reinforces the argument made earlier for a very fast feedforward path based on processing of first spikes. There simply is not enough time to establish a rate at each neuron along the way. Note that the pathway from target onset to RT includes a round-trip from the eyes, to the back of the head where V1 is located, and back to the muscles in the eyes controlling the saccade; this communication alone, even with some myelinated axons, consumes a significant amount of time.

Lee et al. [50] attempted to correlate both NL (the first spike time at V1) and FR (measured at V1) with RT. They found statistically significant correlation between RT and NL and no statistically significant correlation between FR and RT. Hence, after passing through several layers of neurons from V1 to the motor outputs, some echoes of the precisely timed spikes at V1 remain, providing further evidence of a fast feedforward information path based on first spike times (but not firing rates).

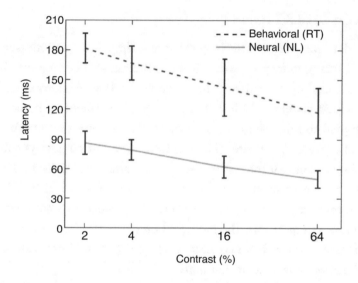

Figure 3.21: Figure 3A from Lee et al. [50], showing both NL and RT as a function of the contrast in the target image. Copyright © 2010 The American Physiological Society.

3.6   OSCILLATIONS

Although we focus on feedforward systems, feedback deserves some discussion. A crucial issue regarding the construction of any large *feedback* system is the presence of some type of synchronizing mechanism. Probably the best evidence of some type of synchronization mechanism comes from observed oscillatory brain activity. An excellent summary of oscillations and synchronization is in Singer [89].

> "... *often, one observes a relation between oscillation frequency and the distance over which synchronization is maintained. Synchronization among remote groups of neurons, or among large assemblies of neurons, tends to occur at oscillations in the theta or beta fre-*

quency range, whereas the highly precise synchronization of local clusters of cells is carried
out by gamma oscillations."

That is, for large assemblies that cover a wide area, the oscillation frequencies are lower, and for smaller local assemblies, the oscillation frequencies are higher.

Two important oscillation frequencies are theta (4–10 Hz) and gamma (40–100 Hz). Recall that spike volleys resulting from saccadic behavior occur at about 4–10 Hz. This is consistent with theta rhythms. The spike coding interval for an individual volley is 10–25 msec, which aligns with the period of gamma oscillations.

3.6.1   THETA OSCILLATIONS

Most "brain wave" data comes from attaching external scalp patches and measuring oscillations at a very coarse scale. Getting more precise data is difficult because collecting it from human subjects involves inserting probes into the subject's brain. However, Raghavachari et al. [83], as part of a study involving epileptic patients, did just that for a group of voluntary human subjects. Having done so, they attempted to find coherence in theta oscillations over the surface area of the cortex. To summarize briefly, "nearby gated sites (20 mm) were often but not always coherent."

Given the observation that there is sometimes coherence over a distance of 20 mm, observe that this covers an area of roughly 400 mm^2. Using data from Figure 3.22, this is a total of 4 x 10^7 neurons that may receive coherent theta oscillations. Forty million synchronized neurons can do a significant amount of computation. The biological presence of coherent oscillations over a large number of neurons is a more-or-less necessary condition for neural network synchronization, so with theta oscillations, we at least clear that hurdle.

It also seems reasonable for synchronized neurons to operate with a period of about 200 msec. If 5 Hz is the input saccade rate, then it makes sense for neurons to operate at roughly the same rate. Internally, the hippocampus also seems to be governed by theta rhythms.

In other words, theta rhythms may synchronize at the individual neuron level. The path between synchronizations contains only a single neuron. The digital logic analogy is a NAND network with a flipflop on the output of every NAND gate. Needless to say, such a logic network would be tightly synchronized.

It is important to note, however, that there are no flipflops or similar storage elements in the neocortex. Rather, inter-neuron "storage" is in the form of delays. If synchronization occurs at individual neuron latencies, it would be relatively easy to establish delay values that yield a synchronized system.

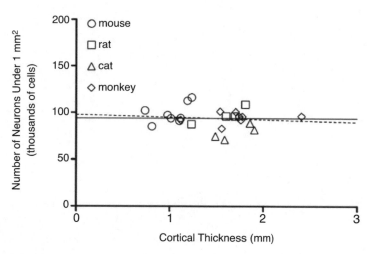

Figure 3.22: Figure 2B from Carlo and Stevens [16]. Carlo and Stevens show neuron densities of 10^5 per mm². Copyright © 2013 National Academy of Sciences. Used with permission.

3.6.2   GAMMA OSCILLATIONS

Vidal et al. [104] conducted a series of experiments with human subjects that required focusing attention to discern similarities in visual bar patterns. Focused attention caused gamma oscillations at 44–66 Hz over a fairly large area in the parietal lobe. The attention task used by Vidal et al. requires 600–700 msec to complete, much longer than the fast feedforward sensory processing driven by saccadic behavior (Section 3.5). If the attention task were done in a feedforward network, the network would be about 60–70 neurons deep, using the same assumptions as in Section 3.5. One may therefore conclude that there is significant feedback processing in the focused attention task.

The gamma frequency oscillations are generated by a mass of interneurons which cause oscillatory inhibition (Cardin et al. 2009 [15], Sohal et al. 2009 [92]). Oscillatory inhibition will essentially enable and disable excitatory neurons in a synchronous way. That is, at the peak of inhibition, excitatory neurons are disabled from firing. At the trough of inhibition, they are enabled.

PART II

Modeling Temporal Neural Networks

CHAPTER 4

Connecting TNNs with Biology

TNNs as a general class are isomorphic to space-time computing systems. Considered abstractly, they exist as computing paradigms completely separate from biological neocortex operation. However, the reality is that we want to construct computing devices that can do the same kinds of things that the neocortex does, so we attempt to replicate the major features of neocortical operation.

It is asserted that the basic paradigms used by the neocortex fit into the broadly defined set of TNNs. In this chapter, basic neocortical operation is shown to be consistent with basic TNN properties. This will be done by referring to material in the previous chapter that summarizes important experimentally-observed properties of the neocortex. Then, in chapters following this one, discussion narrows to a specific TNN model that draws further on experimentally observed properties.

The most basic features underpinning the TNN model come from the neuron doctrine [12]. The neuron doctrine is a set of broadly accepted features concerning the brain's operation. The doctrine identifies neurons as the atomic units of the brain's structure and function. It identifies elemental components of a neuron: body, dendrites, and axons. And, in the "law of polarization," the doctrine states that information is received at dendrites connected to the neuron body and information is transmitted to other neurons via the axon.

Given the neuron doctrine as a starting point, the first few sections of this chapter deal with reconciling the very complex biological neocortex with the relatively simple TNN computational model. This simplification revolves around the coding of information with voltage spikes. This coding is important in itself, and also because it determines important features of neuron-implemented functions. Then, subsequent sections deal with other important TNN modeling assumptions. These include plasticity and training, fault-tolerance, temporal stability, and noise.

4.1 COMMUNICATION VIA VOLTAGE SPIKES

To the neuron doctrine we add the further assumption that all communication of information between neurons *directly related to computation* is performed via voltage spikes. The neuron doctrine states only that there is transmission of information from one neuron to the next, but it is not limited to a specific physical mechanism. As part of this assumption, the voltage levels and other physical characteristics of a spike are not important; only the time at which the spike occurs is important.

"Directly related to computation" is emphasized above to allow for other communication mechanisms, in addition to spikes, that support fault tolerance and temporal stability, for example.

However, our interest is in computation, not biological methods for maintaining fault tolerance and temporal stability (more on fault tolerance and temporal stability in Section 4.8).

The abstraction of voltage spikes to points in time is illustrated in Figure 4.1. Along the top is a train of voltage spikes, with time moving from left to right. As illustrated in the figure, in the baseline TNN model, information is conveyed via spike times, having a timing resolution of ε. The value of ε can be made arbitrarily small, as long as it is not infinitesimal. Given that we are only interested in the presence or absence of a voltage spike, and measure actual time in units of ε, we can abstract the spike train to a stream of ones and zeros to get a binary representation as shown in Figure 4.1 (middle).

With the binary representation, however, some of the visual impact of a voltage spike train is lost. Consequently, an idealized spike train representation (Figure 4.1, bottom) is used in figures throughout this book. It retains the visual appeal of physical spike trains, and it abstracts the voltage spikes to a binary feature in the representation.

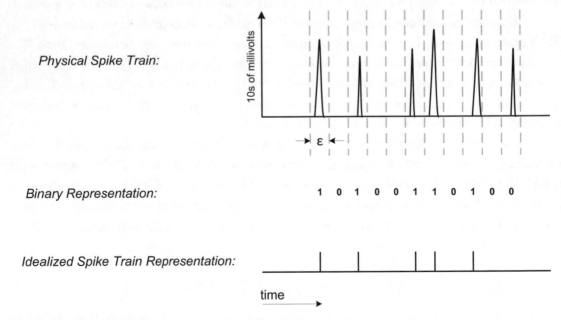

Figure 4.1: The abstraction of voltage spikes belonging to a spike train.

4.2   COLUMNS AND SPIKE BUNDLES

It is assumed that biological neurons are organized into computational groups, or *columns*, that operate in concert. This is in accordance with columnar structures observed in the biological neocortex. It is assumed that the output spikes from a column encode information that is communicated on a bus-like structure. Refer to Figure 4.1. Define the connections from one neuron to the next to be *lines*, with parallel sets of lines organized into *bundles* (analogous to a bus in a conventional digital system).

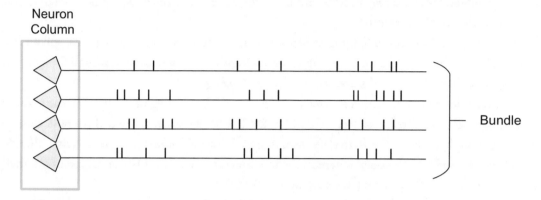

Figure 4.2: A column of neurons produces spikes on parallel lines, or bundles, that encode and communicate information.

4.3   SPIKE SYNCHRONIZATION

It is further assumed that the spike trains belonging to a bundle are separated into waves, with relative quiescence in between. This yields a form of synchronization, and there are at least two ways this type of behavior may be achieved, discussed in the next two subsections.

4.3.1   APERIODIC SYNCHRONIZATION: SACCADES, WHISKS, AND SNIFFS

This type of synchronization appears at sensory interfaces. As discussed in Section 3.5, the senses provide neural stimulation in bursts, for example saccades, whisks, sniffs. These are all lumped together as "saccadic behavior" in the following discussion.

The spike trains invoked by saccadic behavior and their associated processing strongly suggest a special-case synchronizing mechanism. What makes it "special" is that (1) synchronization is external to the cognitive path, and (2) synchronization occurs at irregular time intervals.

Despite the irregular spacing of saccadic spike bursts, they appear to support useful computation. For example, there is good experimental evidence that saccadic bursts are processed in a feedforward manner that enables fast identification and motor tasks; see Section 3.5. Especially note work by Maldonado et al. [61] and Reingold and Stampe [84].

4.3.2   PERIODIC SYNCHRONIZATION

Relying on saccades for synchronization seems plausible only for processing that takes place near the sensory inputs. Saccades may be sufficient for driving a few neuron levels, but not for controlling all the processing in the neocortex.

Assuming neocortical processing is hierarchical, as one moves to higher levels, spike information from ever-broader channels are merged. At a high level, where inputs from different senses come together, there is a special name for the merge: *binding*.

The need for synchronization is often associated with the binding problem [89]. For example, the visual appearance, smell, and feel of a peach are bound together to yield the single concept: "peach." Although each of the individual senses may be synchronized by the external stimulus (saccade, whisk, sniff), at the point where they come together there must be some type of internally generated synchronization that allows binding to take place.

It is also very common for spiking information to be fed back and merged with spiking information that entered the system at a different time. When this occurs, a spike train is essentially merged with a temporally displaced and processed version of itself.

Regardless of whether feedforward binding or feedback are involved, implicit in both is the assumption of a *periodic* synchronizing mechanism that assures that spike information is coherently aligned as it merges. As described earlier, biological mechanisms such as saccades can provide some *aperiodic* synchronization at the sensory periphery, which is adequate for driving feedforward processing in the absence of binding, but saccadic-like phenomena at the periphery are too irregular to provide the periodic synchronization for binding and internal processing of dynamic spike patterns.

Furthermore, such a periodic synchronizing mechanism must cover a large number of neurons and synapses. A single synchronizing mechanism doesn't have to synchronize the entire neocortex or even an entire region, but even if only a single macro-column must be synchronized, it must cover about 10^5 neurons and 10^8 synapses. This fairly large conglomeration of neurons and synapses must then be synchronized to a fairly high precision; say, on the order of a 1 msec skew for pairs of directly communicating neurons.

Inhibitory oscillations (Section 3.6) may offer a way of providing this synchronization. In fact, such oscillations appear to be about the only plausible mechanism that is present. An important point is that although oscillations involving inhibitory neurons may divide processing into periodic windows, oscillations do not provide any kind of static storage analogous to a latch or flipflop in a digital system. Rather, a delay can be viewed as a storage element—much like mercury delay lines were used for storage in some of the earliest computers. Hence, the flow of information, as reflected in spike timing, is not interrupted by the synchronization process (Figure 4.3).

Finally, observe that modulation of the type just described is not confined to temporal spike coding as assumed in this work. If one were to assume rate coding (rate modulation) instead, there must still be some kind carrier oscillation across multiple spike trains in order to assure coordinated neuron processing.

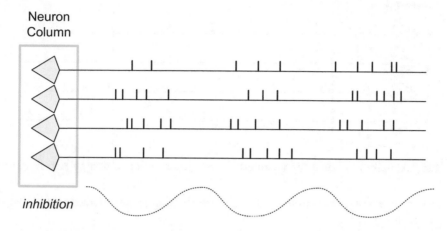

Figure 4.3: Oscillatory inhibition divides spike trains into distinct bursts.

4.4   FIRST SPIKES CARRY INFORMATION

At least near the sensory-motor interface there is strong experimental evidence, summarized in Section 3.5, that the times of the first spikes in a burst are the key to plausible, fast-response information communication and processing. It appears that the first spike in a burst carries sufficient information for accurate processing. Consequently, for modeling purposes, it is assumed that all the spikes in a burst after the first can be ignored. The first spike abstraction is illustrated in Figure 4.4. In this temporal abstraction, the set of parallel first spikes are referred to as a volley, and one may assign values to the spikes in a volley by using the coding method used informally in Section 1.3.2, to be formalized later in Section 6.2.

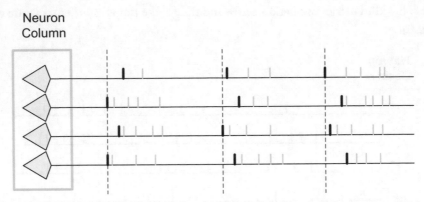

Figure 4.4: The first spikes in bursts are assumed to carry information. Other spikes are ignored.

A summary of the experimental justification for the first spike abstraction follows; details are given throughout Section 3.5.

1. First spike times in response to repeatedly stimulating the same neuron are precise (to within about 1 msec) [60, 79]. Good precision is a requirement if relative spike times are going to be used for coding information.

2. In carefully controlled experiments, human subjects can recognize objects so quickly that neurons along the feedforward pathway only have time to transmit (and operate on) information coded with first spikes [97].

3. More information regarding the stimulus can be extracted from the first spike than from the rate [43, 46, 50].

4. In the pathway between vision and a motor system response (a saccade), there is significant correlation between first spike time at the lowest region in the visual cortex (V1) and the response at the motor output (the initiation of a saccade). This correlation between input and output is maintained despite the several neuron layers that must separate the input from output. There is no such significant correlation between the spike rate and the motor system response [50].

Beyond experimental support, there are additional features and advantages that accrue from using the first spike temporal abstraction. One such feature is that this encoding method is emergent. The information conveyed by a spike on a single axon depends on spike times on the other axons. The complete encoded information on a set of axons emerges from the collection of lines' first spikes, yet any axon considered in isolation offers only minimal information: the presence or absence of a spike on that axon. Furthermore, volleys composed of first spikes are consistent with causal neuron operation; that is, the mere presence of the first spike in a volley initiates neuron processing without the need for any global synchronizing mechanism.

In neuron modeling to follow, only first spikes (per line) will be included in the computational model. Given the assumption that the first spike on each line carries the information, a question immediately arises: why do real neurons in a sensory pathway appear to communicate via spike bursts if the first spike alone carries the information? There are at least two potential answers the author can come up with (the reader is invited to come up with others!).

1. A first-spike coding method is based on an electro-chemical mechanism that causes the more intensely stimulated neurons to spike sooner. Additional spikes may be nothing more than an unavoidable byproduct of quickly producing the first spike. To get a fast first spike, a lot of current must be pumped into the neuron body over a short period of time, and this necessarily will result in a burst of spikes. The effect cannot be confined to a single, isolated, fast spike.

2. Spikes in a burst after the first may provide some non-excitatory functionality. In particular, a burst of spikes may be good for driving faster-responding and sustaining longer-lasting inhibition [45].

4.5   FEEDFORWARD PROCESSING

TNNs are feedforward networks. As shown in Section 2.2, unrolling an arbitrary feedback network to yield a feedforward network does not diminish its computational capability, although it might affect its implementation efficiency.

Going beyond the theoretical argument for studying feedforward TNNs, as described in Chapter 2, there is also significant experimental support for fast feedforward processing in the biological neocortex. This processing enables rapid response, survival-level cognitive skills. We humans, and all mammals, rely on the same low-level cognitive skills for survival as many other animals. Our eyes also saccade in response to an object suddenly appearing. We can very quickly recognize a face, smell smoke, or hear and quickly react to the snap of a twig behind us.

A crucial link between the feedforward survival skills, shared by all the mammals, and the advanced human-only skills is uniformity. In mammals all cognitive skills are implemented via the same underlying neocortical mechanisms. Careful study of the various neocortical regions shows that the neurons that perform the survival-level cognitive skills are configured, arranged, and interoperate in much the same way as the neurons in other regions of the neocortex where the high level cognitive processing takes place. Consequently, one can infer that the basic feedforward paradigm used in the neocortex for low level cognition is also an essential part of higher level cognitive functions performed by the neocortex. Hence, we have a biological argument for studying feedforward networks to go with our previous theoretical argument.

4.6   SIMPLIFICATIONS SUMMARY

Following are the steps that take us from feedback-connected neurons communicating via spike trains to TNNs, i.e., feedforward interconnected neurons that process a single volley at a time, each line in the volley containing at most one spike.

We began by assuming axons bundles are associated with columns, and communication takes place by encoding information across lines in a bundle. Then, we assumed spike trains are divided into synchronized bursts. This is done via inhibition, either as a byproduct of saccades or through oscillatory behavior. The next assumption is that for each line in a bundle, the first spike in a burst carries the information for that line, so the later spikes can be ignored. Finally, after unrolling the single-spike "bursts," we are left with a single volley per neuron per computation. By making experimentally supported assumptions regarding spike encoding in the biological neocortex, we have achieved the relatively simple TNN model.

4.7   PLASTICITY AND TRAINING

In the biological neocortex, virtually every feature of the neocortex has some degree of plasticity. Some features are more adaptable than others. Some plasticity takes place over short time intervals, and other plasticity takes place over longer time intervals, including the time span of fetal and childhood development.

As defined earlier, a space-time implementation is composed of a set of interconnected functions N_i. Consequently, to specify a well-defined space-time system both the interconnection structure and the N_i functions must be determined. As observed earlier, these are not designed in a top-down function-driven manner. Rather, a key premise is that a TNN's implementation is developed through an adaptive training process. The implementation is plastic, and the final function is implied.

In the world of neural networks, plasticity of synaptic weights receives a lot of attention. However, in biology there are other elements besides synapses that are plastic. For example, there is plasticity in axon diameter that can affect spike transmission delays [10]. Importantly, there is also plasticity in connectivity; that is, dendrites can grow, and shrink, and synapses can appear, and disappear, over long time scales [19].

The TNNs of interest in this book are *trained* by emulating their expected operating environment. In particular, a set of pre-determined training input volleys is first applied. Based on these inputs, the training process establishes the values of neuron parameters and synaptic weights, which collectively capture (or learn) important characteristics of input patterns.

Training can go beyond parameters and synaptic weights, however. A training process may be used to arrive at a specific network interconnection pattern, for example, by trying a number of pseudo-random candidates.

For the TNNs discussed in this book, there are two important assumptions:

1. Virtually everything is potentially plastic. As part of the ongoing research effort, TNNs are implemented in a software simulator, so there is a lot of flexibility that can be studied: the neuron configuration, connectivity, parameters, weights, etc. When exploring the design space, it is up to the person running the model simulator to decide which features are plastic and which are not.

2. Training is separated from computing. Eventually, we strive for learning and cognition that are continuous, ongoing, and concurrent processes. However, in the research presented here, as is the case with much of neural network and machine learning research, training is performed separately from, and in advance of, the processing. That is, training first takes place and establishes network and neuron parameters and weights. These become fixed in the network. Then all processing is performed on that fixed network.

Even though training and processing are separated in the model, it is natural to develop them in parallel because they are tightly coupled operationally. Consequently, model development consists of a series of paired training-evaluation runs with different configurations and parameters for each run.

The decomposition into two distinct modes of operation, training and evaluation, makes the problem significantly easier to study. Moreover, it does not preclude the combined, concurrent operation of the two modes as the model is further developed. In the TNN model developed here, both training and evaluation consume time that is linear in the size of the network and input set, with a small difference in constants. Consequently, the author is optimistic that eventually merging the two will be relatively straightforward.

4.8   FAULT TOLERANCE AND TEMPORAL STABILITY

This subsection differs from the others. The others explain why certain biological features suggest *inclusion* of a similar feature in a TNN model. This subsection provides the author's rationale for *excluding* a large number of biological features from the model.

In particular, it is asserted here that a very large amount of the neuron connectivity and functioning is dedicated to fault-tolerance and temporal stability, not basic computation. Going further, it is asserted that the basic computational paradigm in the neocortex is more-or-less independent of features devoted to fault tolerance and temporal stability. Although they may not be entirely independent in reality, assuming so may be a good first order approximation.

Very little is known about relationship between computation and fault tolerance and stability in the neocortex. Consequently, this section is written from the author's perspective and relies primarily on analogies drawn from conventional digital computer design.

4.8.1   INTERWOVEN FAULT TOLERANCE

Figure 4.5 is a highly fault-tolerant digital logic network. It can suffer wire or gate failures, especially transient ones, often concurrently, and still recover correct values as part of the normal computation flow. By today's standards, it is highly fault tolerant.

In this example, we are given the implementation, but what exactly is the implied function? Given only this network, with no documentation, even an experienced logic designer would be flummoxed when faced with this question.

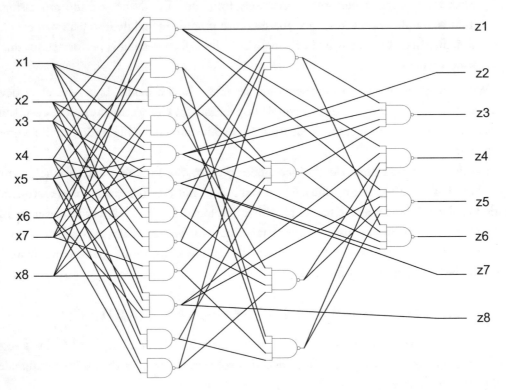

Figure 4.5: What is the function? (Obfuscated version of Figure 3 from [35]).

The function is a half-adder for population coded inputs and outputs. To be specific: inputs x1, x3, x7, and x8 encode input A; inputs x2, x4, x5, and x6 encode input B; outputs z3, z4, z7, and z8 encode the binary sum of A and B, and outputs z1, z2, z7, and z8 encode the carry. With population coding, if three or more of the four input or output lines are the same (0 or 1), then the coded value is a 0 or 1, respectively.

A half adder is a function every beginning computer engineering student learns; in fact, it is usually one of the very first functions to be taught. What makes the above network appear to be so complicated even though the function is elementary? The answer is that it is constructed via inter-woven or quadded logic [80, 99], a fault tolerance technique put forward over 50 years ago. A fault tolerant inter-woven version of any function can be generated in a straightforward way, given the simple non-fault tolerant version.[5] Furthermore, there is no restriction to 4-way inter-weaving as with quadded logic; for higher levels of fault tolerance, higher degrees of inter-weaving can be used.

This example clearly illustrates a computing method composed of functional elements that are not physically separable from the fault tolerance features. The gates for performing the logic function and for performing the fault tolerant function are completely intermingled—they are the same gates. In this case, even though all gates contribute to the final computation, the essence of the paradigm is far simpler than the network suggests. The paradigm is basic Boolean algebra, and the function is a simple half-adder.

The point here is: we do not need to understand how interwoven fault tolerance works to understand the essence of the computational paradigm. Consider this question: would a neocortex composed of less reliable components compute in a fundamentally *better* way? Probably not. Hence, we assume completely reliable components, which frees us from considering all the extra complexity that comes with implementing fault tolerance.

A thesis central to this work is that the neocortex computes with a relatively simple space-time paradigm. Fault-tolerant features analogous to interwoven logic are almost certainly present in the biological neocortex, likely in large quantities, but they are not a required part of the basic computational paradigm. In fact, they may only get in the way of understanding the basic paradigm.

4.8.2   TEMPORAL STABILITY

Any physical computing machine, whether human-constructed or biological will contain timing variations. Every physical device and every physical wire has its own unique makeup that causes its exact timing behavior to differ from all the others, however slightly. To support reliable computation, the effect of these physical variations on timing must be managed in some way, i.e., we want a *temporally stable substrate*.

In conventional computers, we achieve temporal stability by using clocked storage elements (flipflops and latches) and making worst-case timing assumptions. In these clocked synchronous

---

[5] This particular example was intentionally jumbled a little to make it more interesting visually, although the gates and interconnections are the same as the example given and discussed in [39].

systems, physical variations that affect individual device and wire timing do not affect the computed results. We have achieved a temporally stable substrate that supports computation. In the process we have created a system where results of a given step (clock cycle) are independent of the time it takes to compute them.

There is absolutely no evidence of clocked storage elements acting like flipflops in the biological neocortex. Also, neurons communicate with spikes, not levels. Nevertheless, to compute and communicate reliably there must be some degree of temporal stability in the neocortex.

If a model is based on actual time, as is the case with the TNN systems studied in this book, then latencies and delays become an essential part of the model, and they must be used in a manner that is consistent with temporal stability. In particular, when constructing the prototype implementation in this book, certain neuron parameters (time constants for example) and synaptic delay distributions are assumed to be the same across thousands of neurons interconnected by hundreds of thousands of synapses. In systems larger than the ones developed here, there may be many such coordinated groups with each group maintaining its own stable temporal characteristics.

Although we assume in the model that thousands of neurons all have the same parameters, in the biological neocortex, neurons are obviously not all identical. Consequently, an underlying assumption being made in this work is that there is some set of biological mechanisms that work together in a coordinated way, across groups of neurons, to provide a reasonably high level of temporal uniformity.

Note that temporal stability does not necessarily imply a *static* temporal substrate. Although a group of coordinated neurons may have the same parameters and delay distributions at any given time, the parameters and distributions can change, as long as the changes happen in a coordinated way over the group. For example, the frequency of certain inhibitory oscillations may increase or decrease, according to the level of attention. These oscillations may play a key role in maintaining stability as the level of attention changes.

4.8.3   NOISE (OR LACK THEREOF)

It is assumed there is no noise or spike jitter in the prototype TNN. In contrast, it is often observed that there is a significant amount of "noise" in the biological neocortex. For example, noise may take the form of spurious spikes or misbehaving synapses that intermittently go silent.

First of all, knowing as little as we do about the way the neocortex works, it is hard to definitively state that anything is "noise," as opposed to being some very subtle built-in effect. Just because we may not know its cause does not mean it is noise. The book chapter by Thorpe [96] contains interesting discussion of this topic.

In any case, the inclusion of pseudo-random noise in a TNN model may either contribute to computation or degrade it. Obviously, because we are interested in the basic paradigm, we do not want to include performance degrading noise in the model. Consequently, the only noise-like variations in the prototype TNN model are in the form of intentional pseudo-randomness that contributes to the computational method.

4.9   DISCUSSION: RECONCILING BIOLOGICAL COMPLEXITY WITH MODEL SIMPLICITY

At the beginning of Chapter 3, the apparent chasm between the complexity of neuron biology and the simplicity of the TNN model was acknowledged. Because so little is understood about the many interacting biological features, it is extremely difficult to argue why any given complex feature does or does not support a basic cognitive paradigm present in the neocortex. Consequently, in this section, the discussion approaches the topic from a different angle: *if the biological complexity is not there as part of the computational paradigm, then why is it there?*

First, the term "complexity" is a little ambiguous. To be a little clearer, let's say there are two forms of complexity. One is sheer size—the huge numbers of neurons, synapses, axons, and dendrites. By this measure, a system is more complex if it contains more parts. The other form is diversity of components. For example, see [64] for a survey of the huge diversity of inhibitory neurons. The modeling paper by Izhikevich [42] emphasizes the broad diversity of spiking patterns, and there are many other examples. By this measure, a system is more complex if it contains more part types.

There are at least five things that may significantly increase complexity of both part count and part types, but which may not contribute to the basic computational paradigm. These are summarized below.

1. **Spike formation.** There is a lot of apparent complexity in spike trains (both in terms of spike counts and diversity of observed spiking patterns). At the end of Section 4.4, potential reasons for the apparent complexity of spike trains were given. In Sections 4.1–4.5, much of the complexity in spike trains was removed through a sequence of experimentally supported modeling steps, followed by the structural transformation of network unrolling.

2. **Performance enhancement.** This issue was touched upon in Section 1.6. A lot of the complexity in a modern-day superscalar processor is there to improve performance. Analogously, in the neocortex a lot of specialization may be directed at performance optimizations of one kind or another. A performance tweak here, a performance tweak there, may lead to a diversity of neuron types and spiking behavior that is not essential to the paradigm.

3. **Fault tolerance.** This probably accounts for a high percentage the total part count. Redundancy to support fault tolerance is the basis for population coding, for example. This topic was just discussed in Section 4.8.2.

4. **Temporal stability.** This probably accounts for a significant number of neurons and likely leads to some diversity in neuron and synapse types. That is, a significant amount of dynamic tuning is probably required, and this tuning may result in a diversity of part types. Furthermore, maintaining temporal stability probably accounts for a large amount of the feedback that is present in the neocortex, but this is only conjecture.

5. **Noise tolerance.** This is closely related to fault tolerance—in a sense degrading noise appears as transient faults which must be tolerated.

Taken as a whole, all of the above could collectively lead to a very large amount of both size complexity and diversity complexity. Although conceptually separable from the computation, it should be emphasized that the above complexity-increasing features are not physically separable from the computation that is performed.

For example, the inter-woven logic network in Figure 4.5 illustrates a computational paradigm (Boolean logic) that is not physically separable, but is conceptually separable, from features that support a reliable and stable substrate. The author speculates that after we eventually understand the basic paradigm used in the neocortex, we will be able to implement it with biologically plausible fault tolerance methods, analogous to interwoven logic, combined with some form of spike-based population coding to statistically smooth-out individual neuron variations, thereby achieving both the computational capability and the fault tolerance of the neocortex.

4.10   PROTOTYPE ARCHITECTURE OVERVIEW

Having arrived at features of a general TNN model, a specific TNN architecture will be developed in the remainder of this book. It is only one of potentially very many space-time paradigms. This architecture draws heavily from experimental neuroscience research. The TNN architecture then becomes even more fully defined during the development of prototype TNN in Part III of this book. To give the reader some idea of where we are headed, this section contains a brief overview of the architecture direction that will be taken.

At the low level (Figure 4.6), basic processing in the proposed TNN architecture is performed by cognitive columns or computational columns (CCs) that operate on bundles of lines carrying spike volleys. An input bundle may be formed by merging two or bundles from upstream columns. A CC consists of a column of excitatory neurons which are modeled separately and operate in parallel, and two inhibitory columns, one of which acts on the input volley (after merging) and the other which acts on the output volley (prior to downstream merging).

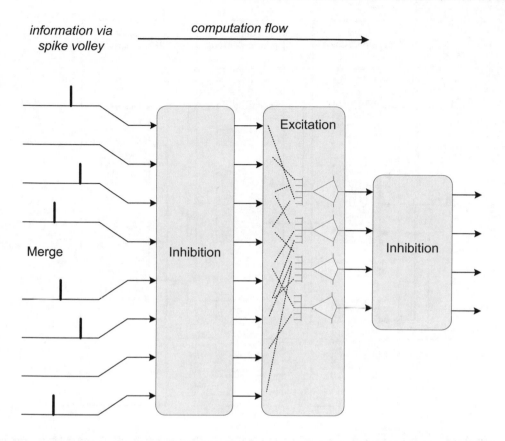

Figure 4.6: A cognitive column (CC) is the basic unit of computation. Merging combines incoming bundles. Excitation computes with first spike information. Inhibition filters by removing certain spikes.

The excitatory columns perform active processing: they take an input volley, perform some operation(s), and yield an output volley. An inhibitory column performs only a filtering function by removing certain spikes as the input volley is passed through to the volley output, but otherwise leaving volley unchanged. Both inhibition and excitation are trained based on input training patterns, although in different ways.

At the higher level (Figure 4.7), groups of cognitive columns are placed in parallel layers that are analogous to macro-columns. A complete feedforward system contains multiple such layers.

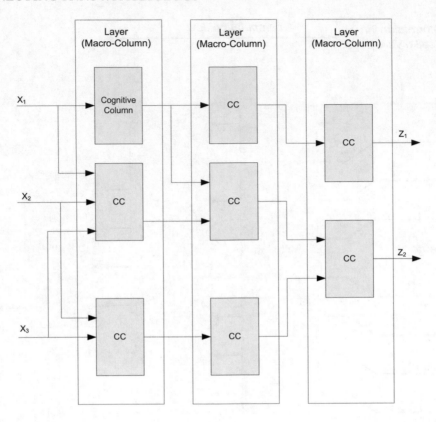

Figure 4.7: A system is composed of a number of cognitive columns, arranged in layers (macro-columns).

CHAPTER 5

Neuron Modeling

Thus far, neuron modeling has been described in general terms. In this chapter, the development of specific spiking neuron models begins. Excitatory neurons are the primary computational elements in the prototype TNN. It is these neurons that observe a volley of spikes at their inputs, process them, and generate output spikes whose timing (and presence) depend on the timing of the input pattern.

Although some of the same model features could be applied to inhibitory neurons, inhibitory functions in the proposed TNN model are performed at a higher level, in bulk. Because inhibition operates at a higher level, the discussion of inhibition is deferred to Chapter 7, where the prototype TNN system architecture is discussed.

5.1 BASIC MODELS

5.1.1 HODGKIN HUXLEY NEURON MODEL

The Hodgkin Huxley (HH) Model [40] is the classic neuron model, for which its developers won the 1963 Nobel Prize in medicine. At the top level, the HH model is based on an RC circuit (Figure 5.1) which models a neuron's membrane potential, v_t in the figure.

From left to right in the figure, the circuit consists of (1) a conductance and reversal voltage (g_t^I and V^I) for inhibitory synapses; the conductance is a function of time, as denoted by the t subscript; (2) a conductance and reversal voltage (g_t^E and V^E) for excitatory synapses; note the difference in polarity for the two reversal voltages; (3) a membrane leakage conductance and rest voltage (G^M and V^{rest}); and (4) a membrane capacitance C^M and associated time-varying membrane potential v_t.

A feature of the full HH model is biological accuracy, e.g., it characterizes the entire dynamic waveform traced by the membrane potential. In the full HH model, the voltage gated ion channels are modeled as a complex set of coupled differential equations. However, as a first order approximation, theoreticians often reduce these to simpler equations of the form: $\tau \, dg_t/dt = -g_t$. This simple first order approximation is used here because it appears adequate for constructing useful computational neurons.

A qualitative discussion of neuron operation was given in Section 1.1. When a spike impinges on an excitatory synapse, the synapse's effects are modeled by the time-varying conductance g_t^E. When the spike reaches the synapse, conductance immediately goes positive by an amount proportional to the synaptic weight, and then begins to decay back toward 0. The presence of a non-zero conductance causes a rise in the voltage v_t across the membrane capacitance C^M. Similar behavior occurs for inhibitory synapses, except when an inhibitory ion gate opens, the voltage v_t is lowered and the offsetting leakage causes it to rise.

Because neurons in the TNN model processes only a single volley at a time (at most one spike per line per volley) the refractory time following an output spike need not be explicitly modeled. Before each input volley, it is essentially assumed that the neuron's membrane potential is quiescent.

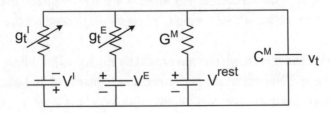

Figure 5.1: RC circuit which is the basis for the HH Model.

5.1.2   DERIVATION OF THE LEAKY INTEGRATE AND FIRE (LIF) MODEL

What follows is a derivation of a spiking neuron model from the RC circuit in Figure 5.1. Many readers may want to skim this subsection; the last paragraph summarizes the results.

The basic HH RC circuit is the starting point for one type of Leaky Integrate and Fire (LIF) neuron model (one of several). The following differential equations characterize membrane potential v_t at time t. Note that in the following, constants are represented in uppercase and time-varying values in lowercase.

(1) $C^M \, dv_t / dt = -G^M (v_t - V^{rest}) - g_t^E(v_t - V^E) - g_t^I(v_t - V^I)$.

(2) $\tau_E \, dg_t^E / dt = -g_t^E$ and $\tau_I \, dg_t^I / dt = -g_t^I$.

If $v_t \geq \theta$ (θ is the *threshold voltage*), the neuron fires a spike and resets to V^{rest}.

These differential equations are modified to streamline their computation. The first modification shifts voltage levels so that $V^{rest} = 0$, this shifts all the other voltage levels accordingly:

$$V^E \leftarrow V^E - V^{rest} \, ; V^I \leftarrow V^I - V^{rest}.$$

Setting $V^{rest} = 0$ in equation (1) and dividing through by C^M yields the following:

$$dv_t / dt = -G^M v_t / C^M - g_t^E(v_t - V^E)/C^M - g_t^I (v_t - V^I)/C^M.$$

Convert to discrete time steps Δt:

$$(v_t - v_{t-1})/\Delta t = -v_{t-1}(G^M + g_t^E + g_t^I)/C^M + (V^E g_t^E + V^I g_t^I)/C^M.$$

As stated in the chapter introduction, we model excitation at the level of individual neurons, and inhibition is modeled at a higher level. Accordingly, the inhibitory terms are not used later, so they are removed at this point in the derivation. However, if one wants to retain them, they are manipulated in much the same way as the excitatory terms (see [91]):

$$(v_t - v_{t-1})/\Delta t = -v_{t-1}(G^M + g_t^E)/\, C^M + (V^E g_t^E)/\, C^M.$$

Define the membrane leakage time constant, $\tau_m = C^M/G^M$, and apply a series of straightforward algebraic operations to yield:

$$(3)\ \ v_t = v_{t-1}[(1- \Delta t/\tau_m) - (g_t^E \Delta t/C^M)] + (V^E\, g_t^E \Delta t/C^M\,).$$

Next, deal with synapses. The synapse conductance g_t^E is modeled as:

$$\tau_E\ dg_t^E\, /dt = -g_t^E.$$

Converting to discrete form, plus simple algebra yields:

$$(4)\ \ g_t^E = g_{t-1}^E\, (1 - \Delta t\, /\tau_E)\, .$$

This equation describes the gradual decay of the excitatory conductance from some initial value. But what value does it decay from? The answer is that in the model there is an instantaneous increase in synaptic conductance in response to an incoming spike. Allowing for multiple excitatory synapses, if an excitatory input spike is received through synapse i at time t, then $s_{ti} = 1$; else $s_{ti} = 0$. Summing over all the input synapses i, at time t, the additional increment in excitatory conductance at time t is $\sum s_{ti}\, w_i\, G_{max}^E$, where w_i is the weight of synapse i (assume $0 \le w_i \le 1$), and G_{max}^E is the maximum initial excitatory conductance.

Add this increment at time step t to the right side of (4):

$$(5)\ \ g_t^E = g_{t-1}^E\, (1- \Delta t\, /\tau_E) + \sum s_{ti}\, w_i\, G_{max}^E.$$

Together, equations (3) and (5) specify an excitatory LIF neuron model to be further simplified below. In an implementation, these two equations yield a *two-stage* model. The synapse stage (equation (5)) defines the state variable g_t^E which holds the synaptic conductance and is subject multiplicative decay. This followed by the membrane stage (equation (3)) with its state variable v_t, the decaying membrane potential, which is a function of g_t^E.

5.1.3   SPIKE RESPONSE MODEL (SRM0)

In the LIF model just derived, the implied subtraction of the embedded term $v_{t-1}\, (g_t^E\, \Delta t/C^M)$ in the expansion of (3) reflects the relatively small effect of synaptic conductance, captured in g_t^E, interacting with the membrane capacitance. Removing this term yields the simplification:

$$(6)\ \ v_t = v_{t-1}(1- \Delta t/\tau_m) + g_t^E\, [V^E\, \Delta t/C^M\,].$$

Ignoring the refractory period, this formulation is functionally equivalent to a version of the widely used SRM0 model: the "zeroth order" version of the more general Spike Response Model (SRM) [30].

In response to an input spike on a line feeding one of the synapses, the SRM first generates a waveform (voltage as a function of time); this waveform is the *spike response*. In biological terms, the spike response is the overall trajectory of the downstream neuron's membrane potential due to the occurrence of a single input spike considered in isolation. For excitatory neurons (as are being discussed here), this is the *excitatory response*.

In the SRM0 (Figure 5.2) individual responses are assumed to be independent and, after multiplying by the synaptic weights, are summed linearly to yield the membrane potential. The neuron emits an output spike at the time the sum of the spike responses first reaches a *threshold* level (denoted as θ). If the sum of the response functions does not reach the threshold value, then there is no output spike.

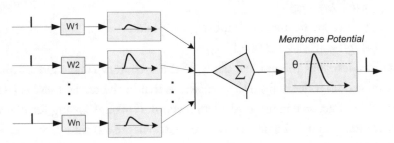

Figure 5.2: The SRM0 model forms a linear sum of weight-multiplied excitatory responses and tests for the threshold θ.

The beauty of the spike response model is that a broad family of spike response functions may be employed in a straightforward way. Two examples are shown in Figure 5.3. On the left is a biexponential response function $E(t) = w_i \, K(e^{-t/\tau_1} - e^{-t/\tau_2})$, where K is a constant. On the right is a piecewise linear approximation of the biexponential. These are but two examples from a very large class of unimodal functions [54].

If we set $K = [V^E \, (G^E_{max}/C^M) \, (\tau_M \, \tau_E) / (\tau_M - \tau_E)]$ for biexponential excitatory responses, then the SRM0 is equivalent to the model given in equations (5) and (6) above. Note: in $E(t)$ given above, $\tau_1 = \tau_M$; $\tau_2 = \tau_E$; w_i = *synapse weight*.

Then the SRM0 with biexponential response functions is defined as:

$$(7) \quad v_t = \sum s_{ti} \, w_i \, K(e^{-(ti-t)/\tau_M} - e^{-(ti-t)/\tau_E}).$$

Having derived the same SRM0 two ways, we also have two corresponding methods for evaluating the membrane potential and trigger firing behavior. In one method, one can start with (5) and (6) and then sequence through a discrete model in Δt time steps to determine when the firing threshold is reached. At each time, step individual input spikes occurring at that time are introduced (if there are any), the state variables are updated, and there is a check to see if the threshold voltage has been reached. The other method based on equation (7) explicitly specifies a set of summed functions to which some type of solver can be applied. The solver finds the first time where the sum is at (or extremely near) the threshold. These two different models yield two correspondingly different simulation methods.

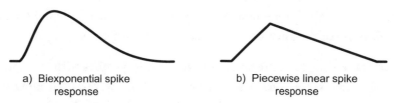

a) Biexponential spike
response

b) Piecewise linear spike
response

Figure 5.3: Two spike response functions.

Which is the approximation? If one approaches neuron modeling from the biological perspective, then a piecewise linear response appears to be a simple approximation for a biexponential spike response. However, it may be that a piecewise linear spike response is a more ideal fit for the fundamental computational paradigm that evolution was driving toward. When viewed from this perspective, it may be that evolution could only approximate a piecewise linear response given the materials and methods that it had at its disposal. Exponential decays are more easily achieved in natural processes than abrupt non-linearities. So, from a computational perspective, perhaps it is the biexponential response that should be viewed as evolution's approximation for what it would have preferred: a piecewise linear response.

The philosophy taken here is to first hypothesize some computational objective produced by a given biological feature, then to construct a model that achieves that objective. The method for achieving the objective does not have to mimic nature, as long as it achieves the hypothesized objective. The implementation of inhibition, to be discussed later in Section 7.7, is a prime example of this approach.

Related discussion on reverse-engineering biological systems (the brain in particular) can be found in the paper by Green [32]. In that work, Green elucidates two views reviewing reverse engineering. One is focused on details and understanding specific parameters; the other is focused on abstraction and understanding higher level functions. These two views are also discussed in Section 1.5, where the distinction is between neuroscience objectives and computer engineering objectives.

5.2   MODELING SYNAPTIC CONNECTIONS

Figure 5.2 illustrates a simple neuron model having synaptic inputs where only a weight multiplication is performed as a spike response is generated. However, to achieve more realistic temporal processing characteristics, delays associated with synaptic inputs should also be modeled.

There are two crucial drivers for the synapse modeling approach to be taken, both discussed in Section 3.4.

1. Synaptic input paths (axon-to-synapse-to-dendrite-to-neuron body) contain delays; these delays are of the same order as neuron processing times.

2. An axon from a single upstream neuron may form synapses with multiple dendrites belonging to the same downstream neuron. In other words, each input connection from one neuron to another may follow multiple paths from the source neuron to the destination neuron body. Each of the multiple paths has an associated synaptic *delay* and *weight*. This is illustrated in Figure 5.4. The individual paths are *simple connections*. The collection of paths from one neuron to another is a *compound connection*. The individual synapses are *simple synapses*. The collection of all the synapses belonging to the same compound connection form a *compound synapse*.

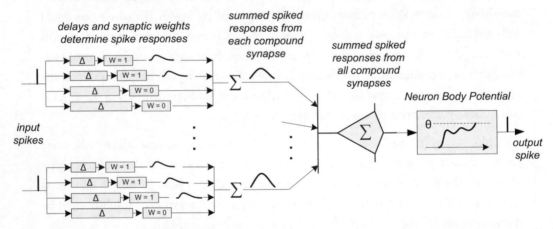

Figure 5.4: Spiking neuron model. Each connection from an upstream neuron connects via multiple paths, each with its own associated weight and delay. The lengths of the delay rectangles indicate the relative magnitudes of the delays.

The simple synaptic weights are typically assigned values in the range [0,1], although this is not a requirement. The weights establish the amplitudes of the individual spike responses. Because a compound connection has multiple synapses, the aggregate weight of a compound synapse may be greater than one. For example, if there are eight paths (and eight simple synapses), the total effective weight of the compound synapse may be as high as eight.

In this model, all delays and weights are associated with the individual simple synapses that make up a compound synapse. The delays may be assigned time values at fixed intervals over a specified range or they may be pseudo-randomly assigned a time according to some distribution. In the current TNN model being developed, the synaptic delays belonging to a compound synapse span a range at fixed time intervals.

It is assumed that the delay paths span some relatively wide distribution. This assumption is based on the experimentally observed distributions of path lengths and spike velocities (Section 3.4). Furthermore, there are a number of other biological features that can significantly affect the neuron-to-neuron delay. For example, these include variations in dendritic and axonic structure, and dendrite and axon diameters. Rather than attempting to explicitly model all such features, it is assumed that all the variables and their experimental ranges combine to yield delays that cover a broad distribution, where "broad" is of the same time scale as neuron computation times.

The model as described thus far is similar to the one used in prior theoretical work [74, 7, 8], as well as more biologically based research [41, 31]. However, the compound synapse training method as proposed later in this book is significantly different from the methods used in this prior work.

The combinations of multipath connections, where the individual paths may have different synaptic weights and delays, define a very large set of potential compound excitatory responses. If there are n simple synapses per compound synapse, and each can have one of m associated weights, then theoretically the compound synapse can implement up to n^m different functions.

For example, if we assume eight simple synapses per compound synapse with delays varying between 0 and 8 msec at 1 msec intervals and weights between 0 and 1 with a resolution of 1/8, (a total of 9 discrete delays and 9 discrete weights), then a single compound synapse can produce 9^9 possible compound excitatory response functions. In contrast, restricting each spike input to a single simple synapse with a fixed delay, as is often done in neuroscience models, yields only 9 different responses (one for each weight).

One of the many modeling challenges is to grapple with this large number of possible compound synapse response functions and come up with a way of assigning specific combinations of synaptic weights such that the neuron as a whole implements a set of useful functions. This challenge is addressed in later sections, and as it turns out, the vast majority of the n^m possibilities are not used.

5.3   EXCITATORY NEURON IMPLEMENTATION

To achieve a simple excitatory neuron implementation, begin with equations (5) and (6), repeated here:

$$(5)\ g_t^E = g_{t-1}^E (1 - \Delta t / \tau_E) + \sum s_{ti}\, w_i\, G_{max}^E;$$
$$(6)\ v_t = v_{t-1}(1 - \Delta t / \tau_m) + g_t^E [V^E\, \Delta t / C^M].$$

Multiply (5) by the constant value $V^E\, \Delta t / C^M$:

$$(7)\ g_t^E [V^E\, \Delta t / C^M] = g_{t-1}^E [V^E\, \Delta t / C^M]\, (1 - \Delta t / \tau_E) + \sum s_{ti}\, w_i\, [V^E\, \Delta t / C^M]\, G_{max}^E.$$

Next, define state variable $v_t^E = g_t^E [V^E\, \Delta t / C^M]$ and meta-weights $w_i^V = w_i\, [V^E\, \Delta t / C^M]\, G_{max}^E$. The yields the following equations, which are directly implemented in Figure 5.5:

$$(8) \; v_t^E = v_{t-1}^E (1 - \Delta t / \tau_E) + \sum s_{ti} \, w_i^V;$$
$$(9) \; v_t = v_{t-1}(1 - \Delta t / \tau_m) + v_t^E.$$

This reduces the evaluation at each time step to the addition of the meta-weight (if there is a spike), followed by two multiplications by constants and a single addition as shown in the block diagram in Figure 5.5. The delays are not explicitly shown, and are assumed to be implemented in the neuron interconnection structure (although they are associated with the synapses).

The illustrated block diagram may be implemented in custom logic, programmable logic, or it may be implemented strictly via software on a general purpose or special purpose programmed processor. A software implementation would use the block diagram as a flow chart, and a hardware implementation would use logic components to implement the blocks.

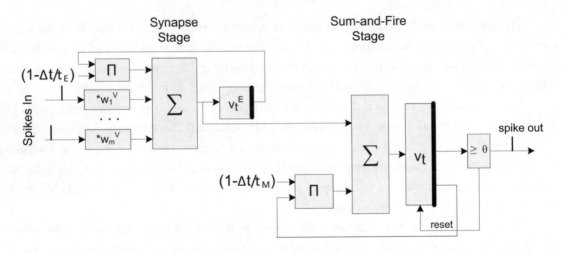

Figure 5.5: Block diagram showing computation performed by spiking neuron model. Synaptic delay elements are not shown.

One feature of the implementation is simple, low precision fixed point arithmetic, although in a software implementation floating point arithmetic may be used for convenience. The execution time of this neuron implementation is dominated by two multiply-adds and a simple test per simulation cycle. Both multiplications have one constant operand (and therefore may be specialized if desired). Note that the apparent weight multiplications at the spike inputs are selection functions (multiplications by 0 or 1, because the spike values are always 0 or 1). There is a fixed computation cost for each spike as it is input to the neuron, but the number of spikes (with sparse coding, especially) will be relatively small, so that computation time is dominated by the two multiply-adds per cycle per neuron.

5.4   THE MENAGERIE OF LIF NEURONS

There are a number of other neuron models that have been proposed and used, which are referred to as "LIF" neurons, sometimes without further explanation. Actually, there are several LIF models, and the differences are significant. They, along with the two models just defined above, are briefly summarized in complexity-order.

As originally defined [94], an LIF model is centered around a membrane potential that decays exponentially (leaks) with some time constant. What typically distinguishes the different LIF models is the modeling of synaptic conductance in response to input spikes. In the following descriptions, the focus is on modelling the synapses and membrane potential. However, in all the models there is also a test for the threshold and the neuron fires if/when the threshold voltage is reached.

5.4.1   SYNAPTIC CONDUCTANCE MODEL

This is the model derived earlier from the high level HH RC circuit. Because it does not use the simplification that leads to the SRM0 model, one might argue that it is more accurate, although it does not appear that this additional accuracy is needed.

Begin with equations (3) and (5) repeated here:

$$(3)\ v_t = v_{t-1}[(1- \Delta t/\tau_{\mathrm{m}}) - (g_t^{\mathrm{E}} \Delta t/\mathrm{C}^{\mathrm{M}})] + (\mathrm{V}^{\mathrm{E}}\ g_t^{\mathrm{E}} \Delta t/\mathrm{C}^{\mathrm{M}});$$

$$(5)\ g_t^{\mathrm{E}} = g_{t-1}^{\mathrm{E}}(1- \Delta t /\tau_{\mathrm{E}}) + \textstyle\sum s_{ti}\ w_i\ \mathrm{G}_{\mathrm{max}}^{\mathrm{E}}.$$

One can use a simplification method similar to the one used in Section 5.4, except the constant $g_t^{\mathrm{E}} \Delta t/\mathrm{C}^{\mathrm{M}}$ must be kept separate from V^{E} because it is used separately in equation (3). Define dimensionless state variable $d_t^{\mathrm{E}} = g_t^{\mathrm{E}} [\Delta t/\mathrm{C}^{\mathrm{M}}]$ and meta-weights $w_i^{\mathrm{D}} = w_i [\Delta t/\mathrm{C}^{\mathrm{M}}]\ \mathrm{G}_{\mathrm{max}}^{\mathrm{E}}$.

Then we arrive at:

$$d_t^{\mathrm{E}} = d_{t-1}^{\mathrm{E}}(1-\Delta t /\tau_{\mathrm{E}}) + \textstyle\sum s_{ti}\ w_i^{\mathrm{D}};$$

$$v_t = v_{t-1}[(1- \Delta t/\tau_{\mathrm{m}}) - d_t^{\mathrm{E}}] + \mathrm{V}^{\mathrm{E}}\ d_t^{\mathrm{E}}.$$

At each time step, this model requires the addition of weighted synapse inputs (if there are any) plus three multiplications (one by a constant) and two additions.

5.4.2   BIEXPONENTIAL SRM0 MODEL

This is the model discussed at some length in Section 5.4. It is the removal of the embedded term $v_{t-1}\ d_t^{\mathrm{E}}$ from the synaptic conduction model just given which leads to additional simplifications.

As noted earlier, the same behavior can be expressed with an SRM0 model, using the biexponential spike response in Figure 5.6a. Each time step, this model requires the addition of weighted synapse inputs (if there are any) plus two multiplications by constants and one addition.

5.4.3 SINGLE STAGE SRM0

This is the LIF neuron model defined originally by Stein [94]. It is essentially an SRM0, which uses the spike response function illustrated in Figure 5.6b.

In this model, all the membrane potential due to an input spike is added in a simple step; i.e., the exponential decay characteristic of synaptic conductance is not modeled (hence, this is a single stage model):

$$v_t = v_{t-1} (1 - \Delta t / \tau_m) + \sum s_{ti} \, w_i \, V^E_{max},$$

where V^E_{max} is a spike's maximum contribution to membrane potential (occurring when $w_i = 1$). As a simplification, one can define meta-weights $w_i \, V^E_{max}$ in a manner similar to that used for the previous neurons. For this model, each time step requires the addition of weighted synapse inputs (if there are any) plus a single multiplication by a constant. This is a very simple model, but simplicity comes at a cost.

In particular, its functionality is less than in the previously described LIF models. As shown by Maass [53], the infinite rising slope (or zero rise time) of the leading edge of the response function means that a single spiking neuron is incapable of implementing some functions that a non-infinite rising slope neuron can implement.

5.4.4 LINEAR LEAK INTEGRATE AND FIRE (LLIF)

The fourth neuron model to be considered is a further simplification where the membrane potential decays linearly rather than exponentially.

This model is the basis for the IBM TrueNorth system [1], where ultra-low energy consumption is a primary design objective. For this model, each time step requires the addition of weighted synapse inputs (if there are any) plus a single subtraction of a constant.

This means that the decay can be performed by subtracting a constant rather than multiplying by a constant. The equation for membrane potential follows:

$$v_t = \max(v_{t-1} - V^L, 0) + \sum s_{ti} \, w_i \, V^E_{max}.$$

V^L is a constant leak value that is repeatedly subtracted from the membrane voltage. This neuron is referred to as a Linear Leak Integrate and Fire (LLIF) neuron. Its excitatory response function for a single input spike is roughly illustrated in Figure 5.6c. It is "rough" because strictly speaking it is not an SRM0; the linear leak is applied at the level of the membrane potential, not the level of individual spike responses.

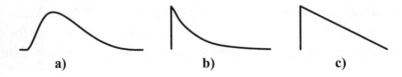

a)   b)   c)

Figure 5.6: Response functions: (a) bi-exponential, (b) Stein's LIF neuron, and (c) LLIF neuron.

5.5   OTHER NEURON MODELS

There are two other neuron models that are sometimes used and warrant further discussion.

5.5.1   ALPHA FUNCTION

Some theoreticians use the alpha function $t \cdot e^{-t/\tau}$ as a spike response instead of the biexponential function. Superficially, the general shape of the alpha function is the same as the more general biexponential (with suitable time constants). And, for modeling purposes, it is only the shape we are interested in. If one implements the spike response by computing the exponential, then this function may be somewhat simpler than the biexponential.

5.5.2   QUADRATIC INTEGRATE-AND-FIRE

Perhaps the best-known example of a quadratic integrate-and-fire neuron model is the one introduced by Eugene Izhikevich [42]. In the Izhikevich model, the membrane potential v is:

$$dv/dt = 0.04 \, v^2 + 5 \, v + 140 - u + I;$$

$$du/dt = a(bv - u);$$

if $v \geq 30$ mv, then $v \leftarrow c$ and $u \leftarrow u + $ d.

where v and u are dimensionless variables,

$a, b, c,$ and d are dimensionless parameters,

t is time, and I is synaptic current.

Rather than checking a firing threshold, the spike occurs naturally as part of the model. The peak of the spike (30 mv in this example) is checked instead. When the spike reaches this peak value, there is a reset.

This model focuses on biologically accurate modeling of the membrane potential, while the synaptic current I is left to the individual modeler. In discussions regarding the accuracy of this model, the emphasis tends to be on the same types of input behavior that experimentalists use for characterizing individual neuron spiking behavior (i.e., artificially applying constant currents—see Section 3.2.2), rather than the types of inputs that naturally occur during normal operation.

In the work in this book where spiking input volleys encode information, the synaptic currents are an essential part of the model, and behavior for constant current inputs is of little concern. Furthermore, it appears that the membrane potential is modeled adequately with simple exponential decay, so a quadratic model is not used here.

Neuromorphic computing—As with many terms "neuromorphic" has evolved to have both a narrow and a broad meaning. Carver Mead originally established the narrow meaning [68]. Mead proposed VLSI analog circuits that communicate via voltage pulses, thereby emulating neural networks. Over the years, the term has become broadened, sometimes to the extent that "neuromorphic" means anything that is neural "inspired," which is about as broad as you can get.

Consider neuromorphic computation as narrowly defined. Does neuromorphic computing make sense? That is, if one were to discover the space-time computational paradigm used in the neocortex, would it be beneficial to implement the paradigm with analog circuits that communicate via voltage pulses that propagate on wires having variable delays? or would it be better to simply simulate neuron operation on a conventional synchronous digital platform, perhaps a special purpose digital platform, or a GPU-like implementation?

The state-of-the-art silicon technology we have today has become finely tailored for synchronous digital operation. This likely makes our current silicon technology unsuitable for direct analog spiking implementations based on precise delays. One example: the main performance objective in current CMOS technology is to minimize delays and latencies. There can be a wide variation in delays and latencies, as long as the worst case is minimized. Furthermore, as digital CMOS technology advances the delay variability between two neighboring transistors tends to increase.

With neuromorphic spike communication and processing, it is the difference between maximum and minimum delays that becomes important, i.e., the variation in delays and latencies should be minimized to allow precise communication via spike timing. Achieving this might require an entirely different way of doing business.

Realistically, at least in the near term, the best implementations will probably emulate spiking neuron behavior with conventional, synchronous CMOS technology; this includes a purely software implementation running on a conventional programmed processor.

5.6 SYNAPTIC PLASTICITY AND TRAINING

Thus far, excitatory neuron models have been described with the implicit assumption that the synaptic weights have already been established. The weights are established via the training of plastic (flexible) synapses. In this work, basic spike timing dependent plasticity (STDP) is assumed—a commonly used first-order model which appears adequate.

Experimental data as collected by Bi and Poo [3] and Markram et al. [62] is summarized in Figure 3.3. This has become an iconic graph that has inspired a number of STDP models, including the one used in this book. STDP modeling is briefly discussed here in a general context. Much of this discussion is based on the definitive STDP modeling paper by Morrison, Diesmann, and Gerstner [70].

Qualitatively, if an input spike (from an upstream neuron) occurs a short time interval *before* an output spike (emitted by the post-synaptic neuron), then the associated synapse should have its weight increased (*potentiated*) because the input spike contributed to the production of the output spike. In contrast, if an input spike occurs *after* the output spike, then its weight should be diminished (*depressed*). This general concept is illustrated graphically in Figure 5.7.

One small difference one might observe when comparing Figure 5.7 with the Bi and Poo data (Figure 3.3) is that at the transition between potentiation and depression in Bi and Poo appears to be more abrupt—it almost looks like an infinite slope. Using a slightly more gradual transition leads to smoother convergence to a final weight; a very abrupt transition may lead to wide swings and instability near the point of convergence.

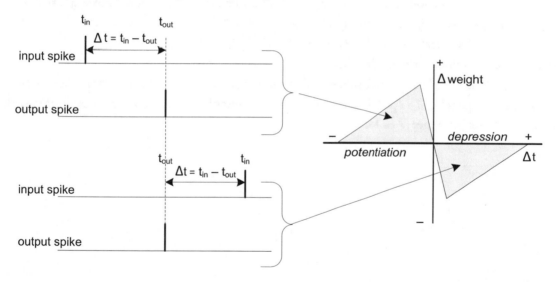

Figure 5.7: Illustration of a typical synaptic update rule. If an input spike precedes the output spike in time, the weight is potentiated (increased) as a function of Δt. If an input spike follows the output spike in time, the weight is depressed (decreased) as a function of Δt.

The specific rules for updating synapse weights due to spike timing relationships are the *synaptic update rules*. As described in [70], synaptic update rules may be divided into two parts, a time dependent part and a weight dependent part. Consequently, the overall update rule for the weight is the product of two functions: F(weight before update) and G(the time difference between the input spike time and the output spike time). The synaptic weights are assumed to be between 0 and 1. Referring to Figure 5.7, define $\Delta t = t_{in} - t_{out}$, where *in* is input spike and *out* is output spike. If Δt is negative, then the positive change in weight (update rule) is

$$\Delta w^+ = F_+(w)\, G_+(\Delta t).$$

If Δt is positive the negative change in weight is

$$\Delta w^- = F_-(w)\, G_-(\Delta t).$$

Both are subject to the constraint that $0 \le w \le 1$, so the final updated weight is

$$w \leftarrow \min(1, w + \Delta w).$$

Next, the F and G functions are defined. First, the F functions. $F_-(w) = \lambda w^\mu$ and $F_+(w) = \lambda(1-w)^\mu$, where $\lambda \ll 1$ is a learning rate, and μ can take on a range of values between 0 and 1. The two end cases are useful in practice as well as for illustrating the general character of the function. If $\mu = 0$, then $F(w) = \lambda$, a constant so the amount of the final update Δw is independent of the value of w before the update. If $\mu = 1$, the amount of the update depends linearly on the value of w before the update. If μ is somewhere between 0 and 1, then the amount of the update depends sublinearly on the value of w before the update.

The two G functions (for + or - values of Δt) may be defined in a number of ways. In Figure 5.7, they are piecewise linear functions. It is also common to use exponential decays:

$$G_+(\Delta t) = e^{-|\Delta t|/\tau+} \text{ and } G_-(\Delta t) = e^{-|\Delta t|/\tau-},$$

or any other functions with similar shape.

Note that F and G are multiplied, and the dimensionless parameter λ is multiplied by their product; consequently, in a specific hardware or software implementation, λ can be associated with the computation of either F or G.

CHAPTER   6

Computing with Excitatory Neurons

Excitatory neurons perform active computation. They take spike volleys as inputs, process them, and generate spike outputs. This chapter develops a formulation for computing with excitatory neurons. Then, in later chapters this formulation (after further refinement) is implemented in the simulator being co-developed with the prototype TNN model.

In this chapter, the TNN neuron model is separated from the physical RC circuit of the previous chapter. Consequently, the model variables become dimensionless, and no longer directly reflect physical quantities as were maintained during model development in Chapter 5. For example, the letters "V" and "v," are re-cycled and are now used for representing a simple *value*, not a *voltage* level as before. One important exception is that actual time continues to be described in units of milliseconds. This is useful for maintaining some contact with temporal aspects of biological neuron operation; it serves as a sanity check of sorts.

In this chapter, the primitive excitatory function, which performs clustering, is described. The sum-and-fire part of the neuron model is very straightforward and is essentially described in equations (5) and (6) in Section 5.1.2. Then, a formal definition of the spike coding method is given before focusing on actual computation, which occupies the remainder of the chapter.

It is the synapses and the training process that get most of the attention in this chapter. Synapse modeling is done in two steps. In the first step (Excitatory Neuron I: Sections 6.4 and 6.5), the multiple synaptic paths connecting two neurons are trained and evaluated separately. Based on empirical analysis of the combined excitatory response functions, the multiple paths are modeled as a single, compound response function to yield Excitatory Neuron II in Section 6.6.

6.1   SINGLE NEURON CLUSTERING

An individual excitatory neuron performs a clustering function. Informally, after unsupervised training with input patterns, the neuron yields similar outputs for similar inputs, where similarity is measured as differences in spike timing. When a single neuron is considered in isolation, many inputs are reduced to a single output, thereby refining, or compressing, similarity information.

Throughout this book, clustering is performed as a two-step process. First, we apply all the input patterns to be clustered; in essence this is unsupervised training. Then in a second step we can re-apply any of the inputs and the network produces the desired cluster (similarity) information in the form of spike outputs.

This clustering function is a primitive building block for the specific TNNs investigated in this book. A description of the primitive function follows, but only after a couple of definitions and a brief meta-computation discussion.

6.1.1 DEFINITIONS

A *feature space* is defined by an ordered set of n features. An object, image, or pattern is represented as a member of the space, a *feature vector*, which contains n elements, each of which indicates some value for the associated feature. For example, a grayscale image containing n pixels can be described with a feature vector of size n, where the value of a feature is the grayscale level of the associated pixel. In this work, spike volleys encode grayscale input feature vectors using a specific method to be described later.

Typically, it is assumed that the values in the feature vector are subject to a *distance measure*, so "closeness" or "similarity" of two feature values can be defined in terms of the difference between the two. The smaller their difference, the closer or more similar they are. The concept of closeness or similarity can be extended to a distance measure for pairs of vectors in the feature space. Consequently, as described in this chapter, a trained excitatory neuron basically measures feature similarity based on distance.

Thus far, the concepts of "similarity" and "distance" have not been precisely specified. And, this will remain the case. As described in Section 2.7.2, we are working in a world of *implied functions* and their *approximations*. In particular, there are two alternatives for describing the model excitatory neuron's basic operation: (1) the exact function implied, by its implementation, with no simpler, more understandable description than the implementation, or (2) an approximate description that is relatively easy to understand and work with, but which is imprecise with respect to the exact function that is implemented.

Because we start with an imprecise description of the distance function we would like to achieve, it follows that some sort of informed trial-and-error process may be the only option for determining the best performing implied distance function. The process proceeds as follows: the designer or modeler establishes the neuron model and parameters; these imply the distance function; the distance function affects evaluation test results; the network designer observes and analyzes the results, and then may make changes to the model. Whether the implied distance function is a good one is ultimately up to the designer of the network, perhaps based on metrics, sometimes subjective metrics involving the observed results.

6.1.2 EXCITATORY NEURON FUNCTION, APPROXIMATE DESCRIPTION

The excitatory neuron model operates in two modes: a *training mode* and an *evaluation mode*.

In the training mode, a set of input patterns are selected as a *training set* for unsupervised learning (in machine learning parlance). In the context of a single neuron's operation embedded in a TNN, it is assumed there is no temporal ordering across the training set, i.e., the patterns in the training set are analogous to unrelated still photos rather than a video stream. Note that from a purely research perspective, this assumption is not as limiting as it might first appear. Following the discussion in Section 2.2, if one were interested in researching video streams as inputs, then in theory one can unroll any video stream into a huge, single-volley pattern. Each training pattern would then be a flat, unrolled video stream, and one can then use multiple such video streams for training. Among the members of such a video stream training set there is no temporal ordering, so the above ordering assumption is satisfied.

In training mode, the members of the training set are applied according to a training regime that is part of the neuron model's specification. The result of the training process is the establishment of synaptic weights for all the neuron's synapses. The values of these weights imply a *center* feature vector. That is, the center corresponds to an input pattern that matches the weights exactly.

After training, in *evaluation mode*, the synaptic weights (and therefore the implied center) are fixed. The neuron's implementation implies its distance function. During evaluation, an arbitrary pattern is applied as a spike volley at the inputs, and the output of the excitatory neuron indicates the distance between the applied pattern and the implied center. Because we are using a spiking neuron, the closer the applied pattern is to the implied center, the earlier in time the neuron will emit an output spike (approximately). Furthermore, for evaluation patterns that exceed a certain distance from the center, no spike will be emitted.

In effect, the implied center plus the implied distance function define a *cluster*, in the machine learning sense. All the patterns that are members of the cluster will produce an output spike; patterns outside the cluster do not. Furthermore, the earlier the spike, the closer the pattern is to the center (approximately). Note that some of the patterns in the training set may not be included in the cluster (i.e., some outliers may be excluded).

Training establishes the cluster center, and evaluation establishes cluster membership based on an implementation-defined distance function. The primitive clustering function just described is the one and only primitive function performed by excitatory neurons in this book.

6.1.3   LOOKING AHEAD

In the remainder of this chapter, information encoding via spikes is first described. Then two excitatory neuron models are developed. The first model (Excitatory Neuron I) trains individual synaptic weights in a more-or-less conventional STDP manner. The second model (Excitatory Neuron II) is based on empirical observation of the first model's operation. The second model employs a higher-level model for compound synapse operation by collapsing a compound synapse into a single, more computationally powerful synapse. This also simplifies the training process considerably. This higher-level model is the one that is used for constructing larger TNN systems later in this book.

6.2   SPIKE CODING

In the TNN model, idealized spikes are transmitted on *lines* that correspond to the axon/synapse/dendrite paths between physical neurons. Multiple lines form a *bundle*; a bundle is analogous to a *bus* in a conventional digital system. Information is coded as a *volley*, which encodes a feature vector wherein each *spike* belonging to the volley is associated with one of the lines in the bundle (so each line carries the value of a specific feature). The abstraction from physical voltage spikes to idealized spikes was illustrated earlier in Figure 4.1.

Because in this book volleys typically represent patterns, and a volley is represented as a vector, the terms "volleys," "vectors," and "patterns" will often be used interchangeably.

6.2.1   VOLLEYS

A spike volley may be denoted either as a *time volley* or as an equivalent *value volley*.

A time volley is a vector wherein each spike represents the relative time t_i that a spike occurs on line i. A time volley is *normalized* if at least one of the $t_i = 0$ and the other spike times are ≥ 0. In a normalized volley, T_{max} is defined to be the latest time at which a spike may occur, so $0 \leq t_i \leq T_{max}$. T_{max} defines the maximum extent of a *temporal frame of reference*, i.e., the time interval over which all spikes in the volley must occur. In this book, volleys may be in either normalized or unnormalized form. However, prior to processing by an excitatory neuron they are always placed in normalized form to establish values within the neuron's local time frame (see Section 2.5).

A time volley is denoted with a t subscript following the vector; for example:

$$A_t = <t_1, t_2, ... t_n>_t.$$

In a single volley at most one spike may appear on a given line, and some lines may have no spike. We use the symbol "-" is another way of denoting "∞", or the absence of a spike on a given line. So, for example, $A_t = <5, - ,6,0>_t$ is a normalized time volley as illustrated in Figure 6.1.

In general, we adopt a convention where the earlier in time that a spike occurs, the higher the value it encodes. This is expressed with an alternate volley representation. A value volley is denoted as $A_v = <v_1, v_2, ... v_n>_v$, $0 \leq v_i \leq V_{max}$. A value volley in vector notation is given a "v" suffix.

There is an inverse linear relationship between normalized time volleys and value volleys: $v_i = \alpha (T_{max} - t_i)$, where α is a proportionality constant: $\alpha = V_{max}/T_{max}$. If there is no spike on line i, the value v_i = "-", indicating the absence of a value. If and only if t_i = "-" then v_i = "-".

Example

The spike volley given in Figure 6.1 is represented as both a value volley and its equivalent time volley. The value volley $X_v = <3, - ,2,8>_v$ is equivalent to the normalized time volley $X_t = <5, - ,6,0>_t$. In the example shown in Figure 6.1, $\alpha = 1$ because $T_{max} = V_{max} = 8$. In general, α may be any positive number. □

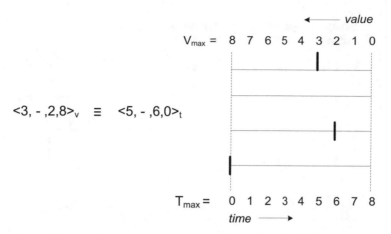

Figure 6.1: An example volley, showing the relationship between values and spike times.

If a time volley is normalized, then the corresponding value volley is also normalized; that is, at least one of the $v_i = V_{max} = \alpha \, T_{max}$. Note that one can easily work interchangeably with either the time volleys or value volleys, perhaps inter-mixing them, because the transformation from one to the other is a trivial, inverse linear mapping. That is, $v_i = \alpha \, (T_{max} - t_i)$ and $t_i = \alpha^{-1} \, (V_{max} - v_i)$.

A generic volley, which may be either a time volley or a value volley, is denoted with no vector suffix; for example: $X = \langle x_1, x_2, \ldots x_n \rangle$. Subscripts distinguish different volleys when needed; for example: X_1, X_2, X_p.

During regular operation, an input volley $X = \langle x_1, x_2, \ldots x_m \rangle, 0 \le x_i \le X_{max}$ is presented to the inputs of a neuron, and the neuron produces an output spike at a point in time that is a function of the input volley (or it produces no spike at all).

When placed in a system (see Figure 4.7), a set of parallel excitatory neurons (a column) produces an output volley $Z = \langle z_1, z_2, \ldots z_n \rangle, 0 \le z_i \le Z_{max}$. In an interconnected system consisting of multiple columns, an output volley from one column is transmitted along a bundle of lines and will typically provide the input volley (or part of an input volley after merging) to a subsequent column.

6.2.2   NONLINEAR MAPPINGS

In the preceding section, and throughout this book, a linear mapping between values and times is used. This is not a model requirement, in general. For example, there can be a logarithmic mapping from values to times. In [41], for example, Hopfield proposes logarithmic mapping.

By using logarithmic mapping, we can map directly from a positional number system to temporal spikes; see Figure 6.2. Each bit of a binary positional number has a weight, which is the base 2 logarithm of its contribution to the total value of the encoded number. Hence, each bit can be ascribed a logarithmic value. Then, one can map these to spike times by using an inverse linear mapping as done previously.

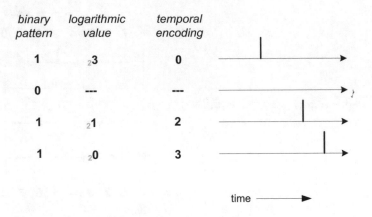

Figure 6.2: Example of logarithmic value-to-time mapping.

6.2.3   DISTANCE FUNCTIONS

A key property of volleys as used in the TNNs envisioned here is that similar input data patterns are represented by similar volleys. However, the implied distance function for determining "similarity" is unconventional and is probably not even a distance metric in the strict sense. Say that normalized training volleys represent a feature space, with a center belonging to the space, and for a given evaluation volley the distance from the center is a function of both the input volley and the center.

A key property of distance is that *a volley's first spike in time (or highest value) contributes more to determining the distance, and therefore feature similarity, than later spikes in time (lower values).* If two different patterns are represented with volleys having their first spikes on the same line, then this is the strongest indicator of similarity between the two volleys. Conversely, if the first spikes in the two volleys are on different lines, the patterns being represented are relatively dissimilar. In general, the second spikes (in time) are not as important as the first spikes for indicating degrees of similarity. This declining importance continues for subsequent spikes, with each subsequent spike being less important than the previous for indicating similarity. At some point, later spikes become irrelevant for indicating similarity (i.e., computing distance).

Example

An often-used distance metric for similarity is Euclidean distance. So, for example, if the center is defined as the feature vector: <8,4,1,7>, then the feature vectors < 7,4,1,8> and < 8,3,2,6> are both the same Euclidean distance from the center. The absolute difference vectors are: <1,0,0,1>, and <0,1,1,0>, respectively, and both lead to Euclidian distances of $\sqrt{2}$. In contrast, with the implied distance function used in this work, the feature vector expressed as a value volley $<7,4,1,8>_V$ is significantly farther from the center than $<8,3,2,6>_V$. $\square$

The reason is that for the implied distance functions use here, not only does the absolute difference of feature values contribute to the distance function, but the individual values of the features also affect the distance. In this example, the feature vector $<7,4,1,8>_V$ differs from the center in the features with the two highest values; the feature vector $<8,3,2,6>_V$ differs from the center in the features with the two lowest values. Therefore, the first vector is (significantly) farther from the center than the second vector, even though Euclidean distance might suggest otherwise.

6.3   PRIOR WORK: RADIAL BASIS FUNCTION (RBF) NEURONS

There is relatively little prior work on TNNs which incorporate multipath compound synapses as used here. Nevertheless, a small amount of important prior work does exist, and discussing it in some detail provides good background for understanding the significance of the training approach and distance function used in the prototype TNN model developed in this book.

One of the conventional machine learning methods for pattern classification is based on Radial Basis Functions (RBFs). RBFs typically determine class membership as a function of a feature vector's Euclidean distance from a defined center.

The RBF concept was transferred to spiking RBF neurons by Natschläger and Ruf [74]. That work was based on earlier observations [41, 31] regarding potential computing capabilities of neurons having multipath compound connections. The structure of the compound synapse model used in [74] is very similar to the one used here. A crucial difference is in the STDP update rules that are used, as will be discussed later.

The main idea behind RBF neurons is that a compound synapse may be treated as a tapped delay line. The simple synaptic weights on a multipath connection are trained in such a way that one out of a range of delays is selected for the path through the compound synapse. During training, all the compound synapses are trained so they select a pattern of delays that determine a center. The key property of the center is that if an input volley is evaluated with inter-spike delays that exactly match the weights defining the center, then the peak membrane potential in the model neuron will be the highest possible. This is an important property that can be used for implementing the implied distance function. All this is illustrated via an extended example.

Example

Figure 6.3 shows a neuron with four compound synaptic inputs. In turn, each compound synaptic input consists of four simple synapses. The four delay values for the four simple synapses are 0, 1, 2, and 3 units of delay (in the figure, the delays are labeled only in the top compound synapse).

Assume that after training with the training set, the center of the cluster is the pattern $<4,1,3,2>_v$. The corresponding time volley is $<0,3,1,2>_t$.

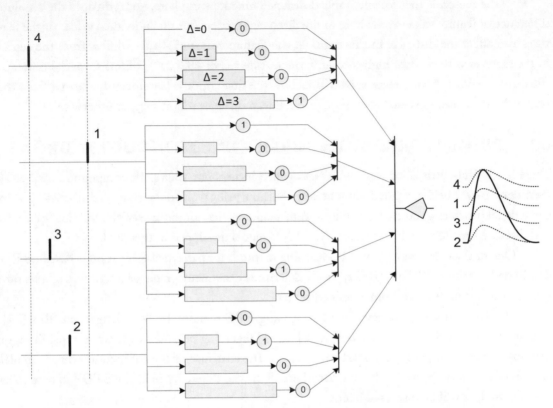

Figure 6.3: Example of an RBF neuron after training for the value volley $<4,1,3,2>_v$.

In the example, after training, a simple synapse in the upper compound synapse has weight = 1 only for the longest delay ($\Delta=3$), and the other weights are 0. Consequently, the response function for the first spike (occurring at $t = 0$) will be shifted by 3 delay units ($t = 3$) before it is added linearly to the membrane potential. The response for the second spike is delayed by 0 time units so it is added in at t = 0 + 3 = 3. Similarly, the response for the third spike is added in at $t = 1 + 2 = 3$; and the response for the fourth spike is added in at $t = 2+1 = 3$. All the responses are aligned in time (at $t = 3$) and are summed at the neuron body. Because all four response functions have the same shape, after the delays align them in time, they will all reach their peak at the same time, and, when added together, their sum will reach its highest possible value. This is illustrated by the waveforms at the outputs in Figure 6.3, where the individual responses (dotted lines) and their sum (solid line) are shown.

Now, consider the distance function. We would like a distance function that generates a spike if an input volley is within a certain distance of the center (and no spike if it is beyond that distance). Furthermore, if there is a spike, the sooner the spike occurs, the closer the input volley is to the center. As discussed earlier, it is the distance function implementation that implies exactly what "sooner" and "closer" mean.

Following is an illustrative example of the way a distance function is implemented in the RBF neuron model. This example clearly illustrates the way model excitatory neurons compute.

Assume the simple response functions have a peak value of 1, and the threshold θ in Figure 6.4 is set to exactly 4. Then the RBF neuron output will spike only if all the responses exactly align in time (as shown), because this is the only way that a membrane potential of 4 will ever be reached. Detecting only an *exact* pattern match with the center is not the objective, however. The objective is to detect not only the center pattern, but *similar* patterns, as well. Consequently, if the threshold is set to a level somewhat lower than 4 (e.g., $\theta = 3.5$), then exact alignment of the responses is not required to trigger an output spike (see Figure 6.4). *Near alignment* of the responses is sufficient, as would be the case with a volley similar to the center volley, but not an exact match. Hence, the value volley $<4,1,2,2>_v$, which is similar to the center $<4,1,3,2>_v$ will reach the threshold of 3.5 and trigger an output spike. Furthermore, in most cases the output spike will occur later than if all the input spikes exactly align (the center). In general, the more similar the input volley matches the trained weights, the sooner the output spike will occur (although this is not guaranteed). Finally, if the input volley is significantly dissimilar to the trained volley ($<4,1,1,3>_v$ in Figure 6.4), the threshold is not reached and there is no output spike.   □

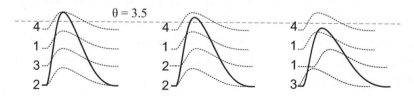

Figure 6.4: An RBF neuron generates an output spike if an input pattern is similar to the trained input pattern, provided the threshold is set somewhat lower than for an exact match. If the input pattern is significantly dissimilar to the trained pattern, then the threshold is not reached and there is no output spike.

In this example RBF neuron, each compound synapse had only one non-zero weight among its simple synapses, so one of the delay paths is selected and the others are de-selected. In general, RBF neurons may have more than one synapse with non-zero weight, but all the non-zero weight synapses should be confined to a narrow region (for more, refer to [74], page 5, first complete paragraph). Hence, the objective in the RBF neuron approach is to train the synapses so that a single synaptic delay, or a narrow band of synaptic delays, is selected (are assigned nonzero weights), and the other synaptic delays on either side of the band are not selected (are assigned zero weights).

To achieve this, the RBF approach uses an STDP update function as shown in Figure 6.5, taken from [75]. The important feature of the rule is that there is only a narrow time region (about 3 msec in this example) where a synapse is potentiated (the value of the update rule is greater than zero). Everywhere else, the update rule causes depression (the update rule is less than zero).

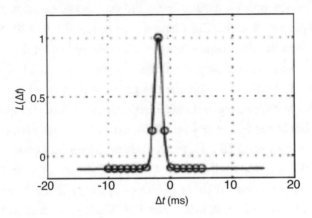

Figure 6.5: RBF update rule. Based on Figure 3a from [75]. Copyright © 1998, Taylor and Francis.

Compare this RBF neuron update rule with conventional update rules (Figure 5.7). Although there are a number of dissimilarities, the crucial difference is the region where the RBF update rule forces synaptic depression, while conventional update rule does just the opposite (potentiation). The reason for the unusual RBF update rule is to isolate a narrow band of nonzero synaptic weights that matches the center pattern, as illustrated in the example of Figure 6.3. The width of the positive region of the update rule is relatively narrow, and determines the width of the band of enabled simple synaptic delays.

The RBF neuron update rule employs a distance function intended to assign close-to-equal importance to all the input spikes denoting a feature vector when determining the distance, independent of their time order. Although it is not Euclidean distance, strictly speaking, it attempts to approximate Euclidean distance in some fashion.

RBF neurons are an illustration of starting with a function (RBF) and then trying to steer the neuron model in a direction that approximates the function. A feedforward network of RBF neurons is a TNN; it follows a space-time paradigm. RBF neurons provide an excellent example of an insightful exploration of the space-time computing paradigm. In this case, the RBF researchers saw an opportunity to perform computation with a TNN, and investigated that opportunity with a non-biological update rule that satisfied their objectives. The resulting RBF neurons and networks yield a very interesting space-time computing method.

6.4   EXCITATORY NEURON I: TRAINING MODE

This is the first part of the definition of the excitatory neuron model used in the remainder of this book. Training establishes synaptic weights and the cluster center is determined by the trained weights.

6.4.1   MODELING EXCITATORY RESPONSE FUNCTIONS

Excitatory responses are first modeled as biexponentials, with a shift toward piecewise linear approximations in later subsections (both biexponential and piecewise linear response functions are shown in Figure 5.3).

In this section, and in the remainder of this book, it is convenient for the weights to directly determine the response function maximum amplitudes; i.e., the weights *are* the maximum amplitudes. A weight of 1 indicates the corresponding response function has a maximum value of 1; a weight of .5 corresponds to a peak amplitude of .5, etc. This was assumed in Section 6.3; it is evident in Figure 6.4, for example.

Aside: Although it is never a significant issue in the work here, for completeness, one can easily determine the value of the parameter-determined constant K in the biexponential excitatory response: $E(t) = w_i K(e^{-t/\tau_M} - e^{-t/\tau_E})$, such that the weight and the maximum response amplitude are equal. The maximum value of the biexponential occurs at time $t_{peak} = (\ln \tau_E /\tau_M)/(1/\tau_M - 1/\tau_E)$ and the peak value v_{peak} is $(e^{-t_{peak}/\tau_M} - e^{-t_{peak}/\tau_E})$. Then, if w_{max} is the maximum weight, setting K = w_{max}/v_{peak} will yield the biexponential where the peak value equals the maximum weight. For example, if τ_E = 5 msec, τ_M = 20 msec, and w_{max} = 1, then t_{peak} = 9.24 msec, and v_{peak} = .472. Hence, if we set K = 1/.472 = 2.12, then the excitatory response function defined as $E(t) = 2.12\, w_i\, (e^{-t/20} - e^{-t/5})$ will have a peak value of 1 when the weight w_i is 1; the peak occurs at t = 9.24 msec.

6.4.2   TRAINING SET

The training input set is supplied by the TNN designer. It is a set of feature vectors, or patterns, taken from a feature space. The choice of members of the training set is part of the application specification. The regime for applying the training set is part of the neuron model definition. For feedforward TNNs, members of the training set are applied in some pseudo-randomized order, with repeats. If the set is large, then one can pseudo-randomize the set once without repeats, and then repeatedly apply the same set.

During training, the model tests for some designer-specified convergence measure. A feature of a good match between the excitatory neuron model and the application is that the weights will eventually converge if enough pseudo-randomly selected training patterns are applied.

6.4.3   STDP UPDATE RULE

The model is based on STDP and the update rule incorporates the piecewise linear function illustrated in Figure 6.6. This function is chosen because the model used here separates processing into distinct and independent volleys were all the spikes in a volley must occur within a coding window (size T_{max} or V_{max}). The spikes in the input volley are processed by the neuron, it produces a spike (or not), and the neuron is then placed in a quiescent state.

In terms of the volley representation, all spikes in a given volley that precede an output spike cause potentiation, and all spikes in the volley that follow an output spike cause depression. Hence, towards the end points, the G update function is set to +1 and -1. Any constant besides 1 could be used, because it is multiplied by a dimensionless constant (λ) in the F update function, anyway (see Section 5.7), and it is the product of the two constants that matters. The slope on either side of $\Delta t = 0$ is to assure that synapses converge to their final weights in a smooth fashion, however this is not necessary.

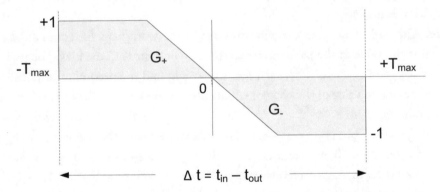

Figure 6.6: Piecewise linear update function G suitable for use when volley processing is divided into abstract time steps.

In the biology, any separation between spikes is dependent on natural delays; in a simulator, however, the volley model with complete separation of volleys is easily implemented, i.e., the TNN completely processes a volley that sweeps through the network before any subsequent volley is applied. Consequently, the G function implemented in the model (and the simulator) is not quite the same as a typical biological model, as illustrated in Figure 5.7. In a biological system where spike volleys are separated only by natural delays, the tapering of G towards zero for larger magnitudes of Δt, as shown in Figure 5.7, but not Figure 6.6, reduces (or eliminates) STDP interference among consecutive spikes. That is, in nature, an input spike always precedes some output spike and it always follows some other, earlier-occurring input spike; so, should it potentiate? or should it depress? The tapering of G in Figure 5.7 answers these questions by allowing a transition between the choice of earlier and later output spikes as time flows past.

For the STDP model used in this work, all that is important is that there is potentiation in the upper left quadrant and depression in the lower right quadrant of the diagram, as appears in experimental data (Figure 3.3). The exact shape of the update curve primarily affects convergence time of the synaptic weights (if there is convergence). That, combined with the volley model which completely separates volleys, leads to the STDP model used here (Figure 6.6).

In one training method, all synaptic weights are initially set at zero. The first volley of the training set is applied to the neuron, and an output spike is forced at the neuron output at a time close to, and before, the maximum time T_{max}. This yields a negative Δt and will cause the weights associated with synapses receiving a spike from the training volley to be incrementally potentiated (Figure 6.6). The process continues for subsequent training volleys. The pseudo-random sequence of training volleys is applied repeatedly until, eventually, the synaptic weights for the given neuron reach a critical point, at which time the neuron will begin to spike on its own according to the neuron model (i.e., the neuron's natural output spike will occur before the forced one). And, at that point, the neuron's synapses will continue learning the volleys being applied and their weights will eventually converge to a statistical steady state. Note that in some cases where the configuration of synaptic delays and maximum weights are too low with respect to the threshold, the neuron may never reach the critical point where it starts to spike on its own. In this case, the weights remain at all 0's.

In most situations, the training process converges to the same set of final weights, regardless of the initial values; that is, they can be initialized to pseudo-random values, or set to any value between 0 and 1. If the weights are set to all 1's (maximum values), then it is not necessary to force output spikes; a neuron will begin spiking on its own, for a typical threshold value.

6.4.4 WEIGHT STABILIZATION

Figure 6.7 is a four input neuron with a training volley shown (both input spikes and resulting output spike). For illustrative purposes, the four spikes in the training volley are arranged from top to bottom, earliest spike time first. Note that in this example, and in others to follow, there is often reference to a single training volley. In practice, there is a set of training volleys. A single training volley is used in examples to clearly establish what the center is; i.e., the training volley itself is the center.

After training, the weights will stabilize as shown. For this input spike pattern, the output spike time is illustrated on the same time scale as the input spikes. At the stable point, all the weights on lines receiving spikes prior to the output spike have been potentiated to their highest values (1), and the weighs on the lines receiving spikes after the output spike are maximally depressed to 0. The stabilization phenomena at work are discussed in [33, 34]. As stated in [34]:

"The dynamical consequence of the asymmetrical, retrograde form of STDP (retrograde because potentiation affects what happened before the post-synaptic spike, thus favoring a "back-in-time" motion), this trend is simplistically explained as follows: for one given input pattern presentation, the input spikes elicit a post-synaptic response, triggering the STDP rule: synapses carrying input spikes just preceding the post-synaptic one are potentiated while later ones are weakened. The next time this input pattern is re-presented, firing threshold will be reached sooner which implies a slight decrease of the post-synaptic spike latency. Consequently the learning process, while depressing some synapses it had previously potentiated, will now reinforce different synapses carrying even earlier spikes than the preceding time. *By iteration, it follows that upon repeated presentation of the same input spike pattern, the post-synaptic spike latency will tend to stabilize at a minimal value while the first synapses become fully potentiated and later ones fully depressed.*" (emphasis added).

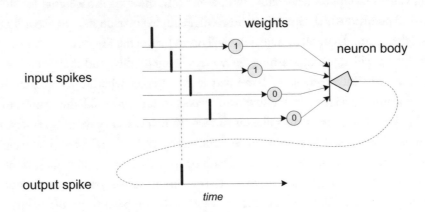

Figure 6.7: Trained neuron after weights have stabilized.

The above example is for simple synapses. Now consider compound synapses, where the delays cover a range of values. For example, Figure 6.8 is a single compound synapse where the delays range in units from 0–3. If a single input spike is applied to the compound synapse, after fanout and delays, the simple synapses see the same input pattern as in Figure 6.7, so naturally, the outcome of training is the same. The weights on the shorter delay paths stabilize at 1, and weights on longer delay paths stabilize at 0.

Finally, consider a neuron with multiple compound inputs (Figure 6.9). An input volley is applied, and each spike in the volley is fanned out in a compound synapse. The output spike reflects the aggregate spiking behavior of all the compound synapses. This output spike then establishes the weight pattern in each of the compound synapses, as described above.

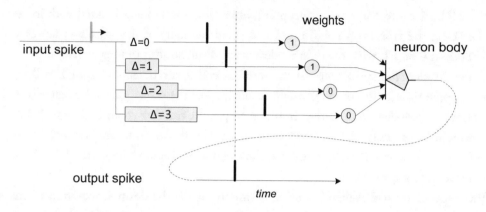

Figure 6.8: Single compound synapse after weights have stabilized.

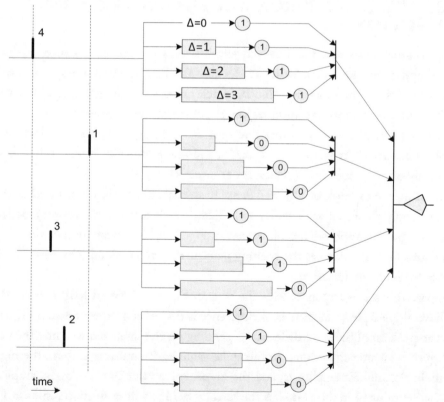

Figure 6.9: Neuron after training for example spike volley.

The bottom line is that every compound synapse stabilizes into a pattern of the form <11...1 f 0...00> where f is a fraction between 0 and 1. That is, the weights associated with shorter delays will all be 1 up to some transition point, after which weights associated with longer delays will all be 0. At the transition point, one (or a small number) of the simple synapses will have weights between 0 and 1, with the weights decreasing from shorter to longer delays.

Depending on the spike times in the training volley, the transition from 1 to 0 may occur early (if the spike time in the training volley occurs late) or the transition may be later (if the spike time in the training volley occurs early). In this example, the training volley is X_v = <4,1,3,2>$_v$. Then, if the threshold is properly chosen (for example, a value of ~ 8), the synaptic weights shown in Figure 6.9 are the stable values. In particular, the weights for input value 4 are <1,1,1,1>, for input value 1 the weights are <1,0,0,0>.

This type of trained weight behavior is important for the desired operation of the model neurons, and this is what sets it apart from other TNNs incorporating multipath connections (see Section 6.3).

6.5   EXCITATORY NEURON I: COMPOUND RESPONSE FUNCTIONS

From the preceding subsection, the idealized weights for a neuron using a more-or-less conventional STDP update rule are of the form <11..1 f00.. 00>. That is, the simple synapses with shorter delays all have weight 1, then there is a threshold delay after which longer delays all have weight 0, although there may be a short transition region where weighs are between 0 and 1.

For the sake of discussion, assume that over the training volley set, the first spike in time (highest value) has a weight vector of <111...11>, and the latest spike (lowest value) has a weight vector with only one or a few leading ones; for example <100...0>.

Continuing the example of Figure 6.9: apply an evaluation volley that matches the training volley. That is, apply the input value volley <4,1,3,2>$_v$, which matches the center exactly. Then the four output compound responses are illustrated in Figure 6.10. The key observation is that if the evaluation pattern exactly matches the training pattern, then the peak amplitudes of the four compound responses more-or-less align.

Compare this set of responses with the ones in Figure 6.3 for an RBF neuron. We observe that both have aligned peak amplitudes. A difference is that with a conventional STDP method as used here, the peak amplitudes are different, depending on the stabilized weights. The peak amplitudes in Figure 6.10 are proportional to the spike value in the training pattern; the highest spike value results in the highest amplitude, and the lowest spike value has the lowest amplitude. Also, the highest spike value (4 in this case) has the least time shift in its evaluation response. In contrast, the RBF neuron exhibits the opposite effect; the highest spike value has the most time shift in its evaluation response.

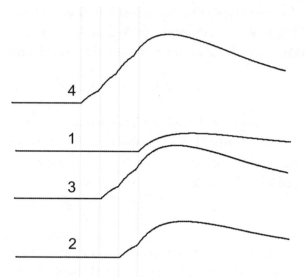

Figure 6.10: Response functions for example value volley after weights and delays have been applied to multipath spiking neurons shown in Figure 6.9.

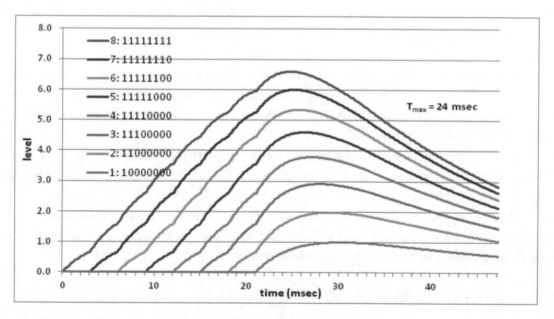

Figure 6.11: Compound response behavior for a range of training values.

To further illustrate the point, Figure 6.11 contains the compound responses for an example with eight paths per connection, with evenly spaced delays. That is, these are the compound responses that are achieved when the spikes having values 1–8 are applied to compound synapses that have been trained for the same values: 1–8. Again, the maximum values of the compound responses nearly align, and the peak amplitude values correlate with the trained weights.

Summary

At this point, the discussion of the first Excitatory Neuron Model ends. One could continue with an implementation by adding a threshold function that operates on the sum of the compound responses. However, the observed compound response behavior suggests a better, higher level model for compound synapses, which is the topic of the next section.

Loose Ends

At the end of Section 5.2, it was shown that if a compound synapse is composed of n simple synapses, and each can have one of m associated weights, then theoretically the compound synapse can implement up to nm different functions.

With the STDP update rule and training method as just described, the synaptic weights after training are always of the form <11..1 f 00.. 00>. The f can be in one of n positions and can take on one of m values. Hence, there are $n \cdot m$ functions. A much smaller number than n^m, but still computationally rich for values of m and n that are each on the order of 8–16.

***

A key property of compound response behavior, is that early spikes, the ones with the higher values, tend to have more influence on the spiking neuron's output. That is, the associated compound response has a higher peak. In a nutshell: *the first spike in a volley identifies the most important feature*. Exactly the property we would like to see.

Briefly, a supporting biological argument for the property just described is that if a spike is first in a volley produced by saccadic behavior (Section 3.5), then the neuron that produces it is the most strongly stimulated of all the neurons producing the volley [96].

This "first spike most important" behavior becomes self-perpetuating in a TNN as envisioned here. The first spike at a neuron's input will also have more time to influence the downstream neurons, thereby passing its higher importance downstream.

6.6   EXCITATORY NEURON MODEL II

The characteristics of compound excitatory responses, as illustrated in Figure 6.11, form the basis for a new more abstract model of synaptic behavior. The abstract model captures the behavior of a compound response without requiring the training and explicit summation of its constituent simple responses. An illustration of the compound synapse model is in Figure 6.12, which plots piecewise linear approximations of the compound responses shown in Figure 6.11. These approximations were determined empirically and are intended to have the same basic characteristics (shapes) as the set of compound excitatory responses, thereby yielding a simpler, more design-friendly computation model.

Excitatory neurons, Model II, are the basic cognitive elements to be used in this, and all following, chapters. In this section, the operation of abstract excitatory neurons is described. As with Model I, the more abstract excitatory neuron model has both training and evaluation modes. First the overall neuron model is described, followed by descriptions of evaluation and training modes.

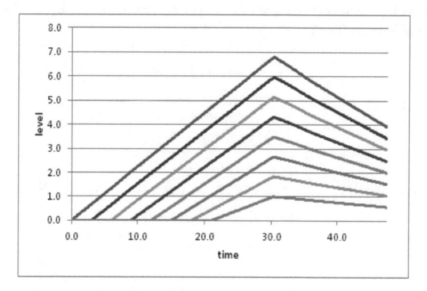

Figure 6.12: Example of a piecewise linear abstract model for compound response behavior.

6.6.1   NEURON MODEL DERIVATION

An abstract excitatory neuron has m input lines, a single output line j, and a *weight* associated with each input-output pair ij. In this model, there is at most a single input connection from any given upstream neuron; all the compound paths in the spiking neuron model given in Sections 6.4 and 6.5 are now modeled with a single path having a single weight that defines a compound response. Because all the simple synapses belonging to the same compound synapse are condensed to a single path, the weight in the abstract model is a single value that is the sum of all the simple weights.

For neuron j, define a compound weight vector $W_j = <w_{1j}, w_{2j}, ... w_{nj}>$, $0 \le w_{ij} \le W_{max}$. Because the weights for simple synapses are typically between 0 and 1, it is convenient to make W_{max} for compound synapses equal to the sum of the maximum weights of the simple synapses. In the example to follow, there are four simple synapses per compound synapse, so in that case $W_{max} = 4$.

As specified earlier, a neuron's inputs consist of normalized volleys with spike times between 0 and T_{max}. Each neuron produces a single output spike that can take on a range of values that depends on a number of factors to be discussed later. T_{max} and W_{max} are related via the scale factor $\gamma = T_{max}/W_{max}$. That is, γ is the ratio of the maximum range of input values to the maximum range of weights. Conceptually, the full coding window (T_{max}) is divided into segments of size γ.

The derivation of the higher-level excitatory neuron model is easiest to explain via an example. Figure 6.13a is a copy of Figure 6.9 given earlier, with relative spike times given instead of relative values. Figure 6.13b shows response functions for the four individual simple synapses making up any of the compound synapses. The simple response functions are piecewise linear with rise time T_r and fall time $T_f > T_r$.

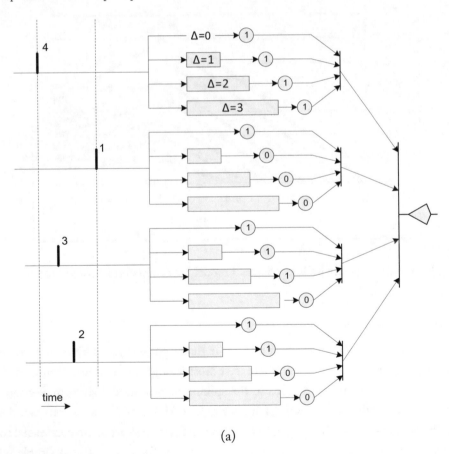

(a)

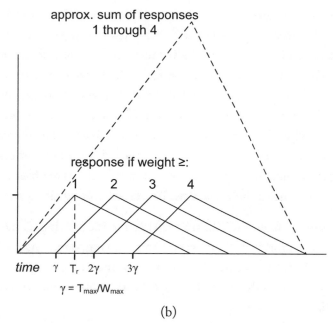

(b)

Figure 6.13: (a) A neuron with trained compound synapses. (b) Each compound synapse generates a group of response functions.

As observed above, we add the weights of the simple synapses forming a compound synapse and assign that weight to the compound synapse. For example, in the top compound synapse, the weight is 4, in the second from the top, the weight is 1, etc.

For a given weight, the set of active simple response functions can be determined from Figure 6.13b, Then these simple response functions are added to yield the response for the compound synapse. However, in this work, we approximate the compound synapse responses for the sums of the simple responses. In particular, we use a simple piecewise linear approximation of the sum, that starts at the same time as the summed simple responses, peaks at the same time, and ends at the same time. The peak amplitude is the weight of the compound synapse.

For example, the top compound synapse has weight = 4, so all four of the response functions are added. These correspond to the four simple synapses, each of which is weight 1. The compound response is the sum of the simple response functions. The approximation for this response is shown in the figure with a dashed line. For the compound synapse second from the top, only the response function labeled 1 in Figure 6.13b is used (because it corresponds to the only simple synapse with weight 1).

Following is the way the approximate compound responses are derived.

1. **Start Time**—Because we are dealing with normalized time, all the compound responses start at t = 0.

2. **Peak Time**—It is assumed that the peak for the compound response is reached at the same time as the latest simple response function reaches its peak. In the example of the top compound synapse, this is the time that the fourth response function peaks. This assumption holds as long as T_r, the rise time for a simple response, is greater than or equal to γ. (Refer to Figure 6.13b.) Observe that if $T_r \geq \gamma$, then response function 2 begins to rise before function 1 begins to fall. Because $T_f > T_r$, the rise in function 2 will more than offset the fall in function 1, so the sum will continue rising. The same can be said about all subsequent response functions except the last. When the last response function peaks and starts to fall, there are no more response functions to offset it. In practice the condition $T_r \geq \gamma$ normally holds, and can be checked during model simulation.

Referring again to Figure 6.13b, we see that for the top input, the fourth simple response peaks at time $3\gamma + T_r$. If we consider the compound synapse third from the top, the weight is 3. In this case, the peak time of the sum corresponds to the peak of response function 3. In this case, it is $2\gamma + T_r$. We observe that in general, the peak occurs is at time $(w-1)\,\gamma + T_r$.

Finally, in the approximation, the peak value is assumed to be the compound weight. In the example shown for the top compound synapse, this is 4.

3. **Stop Time**—the time when the approximate compound response falls back to 0, corresponds to the stop time of the latest simple response. This is one fall time, T_f, beyond the peak, i.e., at time $(w-1)\,\gamma + T_r + T_f$.

The approximate response function $E_i(t,w_i)$ for neuron i having compound weight w_i is summarized in Table 6.1 in terms the scale factor γ and two additional parameters a rise time T_r and a fall time T_f. In the description to follow, T_f is specified directly. Illustration of the piecewise linear functions is in Figure 6.14.

Table 6.1: Definition of response function	
$t < 0$	$E_i = 0$
$0 \leq t \leq \gamma(w_i - 1) + T_r$	E_i rises monotonically until it reaches a maximum value of wi at $t = \gamma(w_i - 1) + T_r$
$\gamma(w_i - 1) + T_r \leq t \leq \gamma(w_i - 1) + T_r + T_f$	E_i falls monotonically until it reaches 0 at $t = \gamma(w_i - 1) + T_r + T_f$
$\gamma(w_i - 1) + T_r + T_f < t$	$E_i = 0$
Note that the above derivation always uses weights that are integer values. In general, weights can be any fixed precision number. In this case, linear interpolation occurs naturally with the given response functions.	

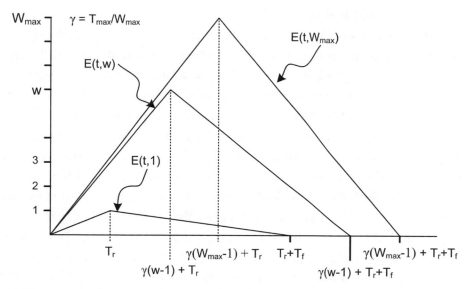

Figure 6.14: Parameterized, piecewise linear compound excitatory responses.

6.6.2   TRAINING MODE

The objective of training is to define a set of weights, and consequently a set of piecewise linear E_i functions as just described. To accomplish this, the average relative times ti in the training patterns are essentially translated into the aggregate weights w_i for each associated line.

The training process itself does not involve repeated application of training patterns with incremental weight changes as directed by conventional STDP. Rather, it is assumed that the spike times for the individual lines in the volley will cause the weights to converge to match the mean input spike time; this was illustrated earlier in Figure 6.8. Accordingly, the mean spike time, transformed by the relationship given earlier yields the trained weight $w_i = (T_{max} - \bar{t}_i) / \gamma$, where $\bar{t}_i$ is the mean spike time for line i.

Briefly, biological STDP is essentially an averaging process that requires a minimum amount of intermediate state. The state, and the way it is maintained, are most likely determined by biological constraints that are not present when constructing a computation model as is being done here. Consequently, weight training in the model is reduced to a conventional averaging process, using the arithmetic mean. That is, rather than having weights slowly converge to the mean, the model simply calculates the mean and is done with it. *This yields simple, localized training with linear training times.*

6.6.3 EVALUATION MODE

After an excitatory neuron has been trained and weights have been assigned, the set of response functions $<E_i(t,w_i)>, 1 \leq i \leq n$ remain fixed as input patterns are evaluated. Evaluation proceeds as follows. An input volley is applied to a trained neuron as a time volley X_t. That is, spike i occurs at the associated time t_i. This causes the response functions as shown in Figure 6.14 to shift by time t_i.

During evaluation, if the relative time of an evaluation spike exactly matches the time of the cluster center (as encoded in the weights), then $t_i = T_{max} - \gamma\, w_i$, and the corresponding response function $E_i(t - t_i, w_i)$ reaches its maximum value at $T_r + T_{max} - \gamma$. That is, if the evaluation pattern perfectly matches the cluster center, the maximum value for all inputs reach their maxima at the same time. Because all the individual weight functions align with their maxima at this time, their sum must also reach its maximum possible at this time. This is illustrated in Figure 6.15.

After shifting, response functions $E_i(t - t_i, w_i)$ are summed linearly. (This sum essentially models the neuron membrane potential), so let $P(t) = \sum_1^n E_i(t - t_i, w_i)$. The sum $P(t)$ is then evaluated to determine the value of t for which its value first crosses the threshold θ. The least value of t for which the linear sum equals or exceeds threshold θ is the output spike time, z. If the sum never equals or exceeds θ then z is null (no spike). Stated concisely:

$$z = min\ t \mid P(t) \geq \theta$$

if there is no such t, z is assigned the null value.

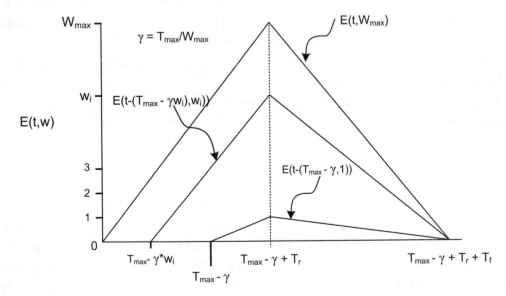

Figure 6.15: If $t_i = \gamma\, w_i$, for all i, the weight functions reach their maximum values at the same time $(T_{max} - \gamma) + T_r$.

Example

Say the weights for a single neuron after training are W = <8,2,6,4>, where W_{max} = 8. Assume T_{max} = 16, so γ = 2. Given W, one can form the four E_i functions. After setting T_r = 10 and T_f = 40, the four E_i are shown in Figure 6.16. Using the properties of the E_i given earlier, E_1 is 0 for values of t less than 0. For values of t between 0 and $T_r + \gamma(w_1 - 1)$ = 10+2(8-1) = 24, the function rises linearly until it reaches a maximum value of 8. Then it falls linearly until $t = T_f + T_r + \gamma(w_i - 1)$ = 64. Similarly, E_2 rises linearly to a maximum of 2 when t = 10 + 2(2-1) = 12. It then falls linearly until it reaches 0 when t = 52. The responses E_3 and E_4 are also shown in the figure.

Continue the example by applying input volley X = <0,12,4,8>$_t$; that is, an input volley that is perfectly matched with the center. Then, the E_i are shifted by the values t_i. The shift amount for E_1 is 0. Similarly, for E_2, E_3, and E_4, the shift amounts are 12, 4, and 8, respectively. This yields the E_i functions shown in Figure 6.17. The sum of the E_i is also given in the figure. Say the threshold θ = 19. Then the function P(t) crosses the threshold at $t \approx 23$.

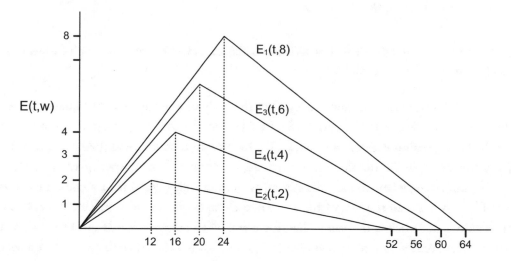

Figure 6.16: Example E_i functions for weights 2, 4, 6, 8.

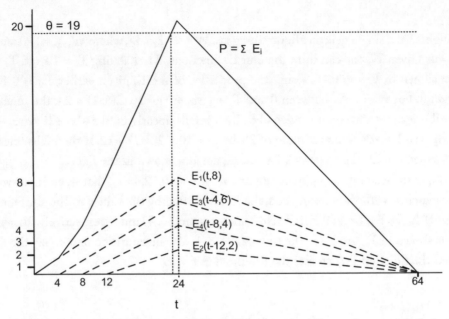

Figure 6.17: Example shifted E_i for T = <0,12,4,8>. The individual E_i are shown with dashed lines, their sum (function $P(t)$) is a solid line.

Consider other example input volleys. For the evaluation volley X = <0,14,6,6>$_t$, the weight functions and their sum are in Figure 6.18a. In this case, the sum crosses the threshold at t = 26. Finally, if the evaluation volley X = <0,14,12,4>$_t$, the weight functions and their sum are in Figure 6.18b. In this example the threshold is not crossed, so the there is no output spike. □

To summarize, when all the spikes in an evaluation volley *match* the trained center weights, i.e., $t_i = T_{max} - \gamma w_i$, then the sum of the associated E_i functions reaches a maximum peak value. If the threshold is at or below this value, then there will be an output spike. If the t_i are *close* to their weight-matched values[6] but not always identical (as in Figure 6.18a), then the sum of the associated E_i will not go as high as the peak value, but it will still reach the threshold. Finally, if the t_i are farther away from their weight-matched values, then the threshold may not be reached and there is no output spike (as in Figure 6.18b). In general, the closer the input volley is to the weight-matched values, the more likely it is to cross the threshold, and the lower the value of $P(t)$ when it does cross.

---

[6] Because of the relationship between time and weight, "weight-matched" and "time-matched" are synonymous.

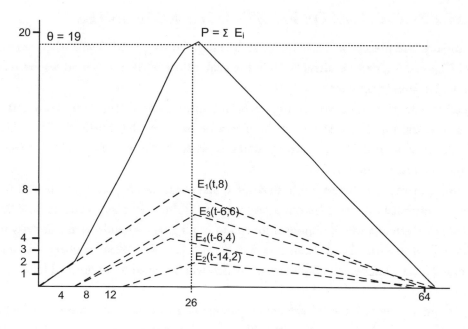

(a) P(t), when X = <0,14,6,6>$_t$

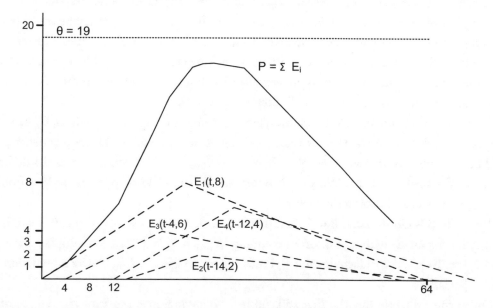

(b) P(t), when X = <0,14,12,4>$_t$

Figure 6.18: Example shifted E_i functions where P(t) crosses the threshold (a), and does not cross the threshold (b). The individual E_i functions are shown with dashed lines, their sum (P) is a solid line.

6.7   ATTENUATION OF EXCITATORY RESPONSES

If one compares the piecewise linear response functions in Figure 6.15 with the compound responses in Figure 6.11, there is attenuation in the peak values of the compound responses, but not in the piecewise linear responses.

Qualitatively, the responses in Figure 6.11 appear compressed, with more compression happening near the top. For example, in the piecewise linear model, a weight of 8 yields a peak weight function value of 8. With the compound responses shown in in Figure 6.11, a weight of 8 corresponds to a maximum amplitude of about 6.6.

In effect, there is an attenuation from 8 to 6.6 that occurs as a byproduct of the shifting of the simple responses that form the compound response. In addition, it is important to note that there may be biological attenuation of responses for spikes impinging on dendrites farther way from the neuron body [93]. As has already been stated in Section 3.4, the collective effects of attenuation will be modeled via a single neuron parameter, which will be tuned to an optimal level during the design process.

Consequently, we will use an attenuation parameter β to achieve a "compressing" attenuation effect. The only wrinkle is that attenuation will be done by pushing up from the bottom. That is, a response reflecting the maximum weight W_{max} will be unchanged. However, responses reflecting smaller weights will have their peak values move up slightly, i.e., closer to the maximum. This yields leads a very simple single-parameter attenuation model, and it also leads to threshold levels that are relatively uniform over a wide range of attenuation. However, modeling this way makes "attenuation" a misnomer because waveforms are actually raised in the model. But, if one also raises the threshold, the net effect is the same as with literal attenuation.

Attenuation is implemented by multiplying the excitatory response functions, E_i, by a value that depends on the E_i function's peak value (and therefore its weight) and the parameter β. Let $p(w_i, \beta)$ be the attenuation multiplier: $p(w_i, \beta) = 1 + (W_{max} - w_i)^* \beta$. $W_{max} - w_i$ is the difference between the E_i's peak value and W_{max}. This difference is multiplied by β so the farther from the peak, the greater the attenuation multiplier.

Typically, β is close to 0. If $\beta = 0$, then there is no attenuation because $p = 1$. If $\beta = .05$, then the multiplier for an E_i with a peak of 7 (assuming $W_{max} = 8$) is 1.05. The multiplier for a peak of 6 is 1.1, etc. Figure 6.19 illustrates the E_i functions for weights 1, 2,..., 8 after attenuation with parameter $\beta = .05$.

Given the principle that the first spike(s) in a volley indicates the strongest, i.e., most important, feature, we can control the degree of "relative importance" via attenuation. The response function peaks provide a de facto way of ranking importance. If $\beta = 0$, then the peaks of the E_i functions are linearly spaced. Increasing attenuation above 0 brings the peaks closer together. This essentially brings their relative importance closer together. Thinking of attenuation as turning a knob that controls relative "importance" provides a useful intuitive metaphor when networks are being designed.

Providing for attenuation extends the unimodal E_i functions given earlier. The E_i functions as previously defined become the special case where the attenuation multiplier, p = 1. The revised E_i functions in Table 6.2 incorporate attenuation as just described.

Table 6.2: Revised response functions to account for attenuation	
$t < 0$	$E_i = 0$
$0 \le t \le \gamma(w_i - 1) + T_r$	E_i rises monotonically until it reaches a maximum value of $p(w_i, \beta) * w_i$ at $t = \gamma(w_i - 1) + T_r$
$\gamma(w_i - 1) + T_r \le t \le \gamma(w_i - 1) + T_r + T_f$	E_i falls monotonically until it reaches 0 at $t = \gamma(w_i - 1) + T_r + T_f$
$\gamma(w_i - 1) + T_r + T_f < t$	$E_i = 0$

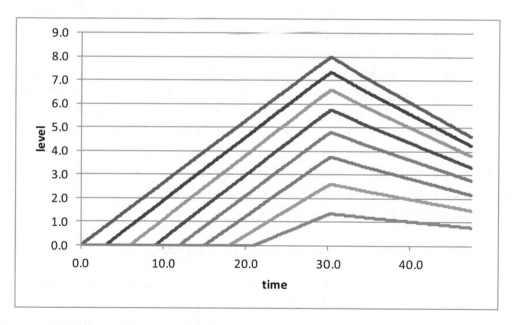

Figure 6.19: E_i functions after attenuation has been applied; β = .05.

6.8 THRESHOLD DETECTION

Thus far, the method for forming the E_i functions has been described, and the threshold for generating an output spike has been defined. What remains is a simple, implementable method for computing the output spike time (value), given the response functions and an input volley.

As discussed in Section 4.1, actual time is a numerical scale divided into increments of size ε. Generally speaking, the time unit ε is determined by the computational resolution required, relative to other time parameters in the model, T_r and T_f in the piecewise linear model. For example, if T_r and T_f are on the order of a few msecs to a few tens of msecs, then choosing $\varepsilon = .1$ msec yields good precision for the computation of the output spike times.

A straightforward evaluation method is to simply step through time incrementally, one ε time unit per step, calculating the function $P(t)$. If the threshold is first crossed at time t, then the output spike time is t. This method works well with biexponential excitatory response functions where the conductance time constants (τ_E) are the same for all synapses. In that case, the exponential decay of the individual synapses summing to $P(t)$ can be implemented in the model as a single decay (Section 5.4). This is not the case with piecewise linear E_i functions, however, so the E_i for each synapse would have to be tracked separately to determine $P(t)$, leading to significantly increased computation.

A better method for detecting the threshold with piecewise linear E_i functions is based on the observation that their sum, $P(t)$, is always linear between breaks in linearity in any of the component E_i. So, after an evaluation volley has been applied, one can identify in advance all of the times where any of the individual E_i functions contains a break in linearity. These are the points (1) where an E_i rise time begins, (2) where an E_i reaches its peak, and (3) where an E_i fall time ends, although in practice only the first two make a difference. Given the nonlinear breakpoints for all the E_i functions, one need only evaluate the function $P(t)$ at the breakpoints, in increasing time order. If at breakpoint i, the threshold is first exceeded, then the threshold spike time must occur between breakpoint i-1 and breakpoint i. Solving a simple linear equation (because $P(t)$ is linear between consecutive breakpoints) will yield the point at which the threshold is crossed. Other, more optimal methods based on bisection may also be implemented.

6.9   EXCITATORY NEURON MODEL II SUMMARY

The neurons that make up an excitatory column are fed by a set of compound synapses, where each compound synapse has an associated weight. The weight defines a compound response function, Ei, based on the parameters given in Table 6.3. The abstract excitatory neuron model in schematic form is illustrated in Figure 6.20.

T_r and T_f are defined to have the relationship $T_f = \rho \cdot T_r$. So, when the parameter T_r is assigned a value, T_f is automatically assigned a value as well.

Table 6.3: Parameters for excitatory neuron model II	
T_r	E_i rise time when $w = 1$
ρ	$T_f = \rho\, T_r$; E_i fall time when $w = 1$
T_{max}, V_{max}	maximum time in a normalized time volley; maximum value in a normalized value volley $V_{max} = \alpha T_{max}$
W_{max}	maximum weight of a compound synapse
$W_{max} = T_{max}/\gamma$	
β	attenuation parameter for E_i multiplier: $\beta^* (W_{max} - w) + 1$
θ	threshold

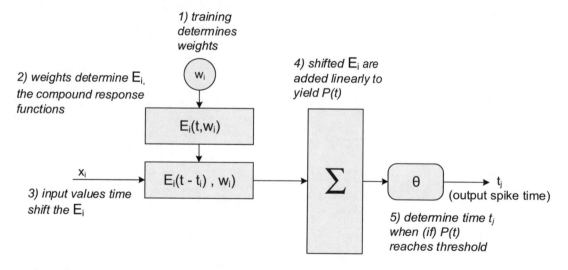

Figure 6.20: Schematic diagram of excitatory neuron operation.

CHAPTER 7

System Architecture

In this chapter, the overall architecture of a prototype TNN is described. This architecture will be used for the extended design study in Chapter 9. The design to be undertaken in that chapter is a clustering TNN for a long-standing machine learning dataset. Although placed in separate chapters, the ideas in this chapter and experimental results from the extended design study Chapter 9 are being co-developed as an iterative process.

First, a section containing a system architecture overview is given, followed by a series of sections containing descriptions of the major components in greater detail.

7.1 OVERVIEW

Figure 7.1 is the block diagram of the prototype TNN architecture. The architecture consists of a number of layers, each containing a number of parallel *computational* or *cognitive columns* (CCs) where each CC contains a number of parallel excitatory neurons combined with bulk inhibition (shown in Figure 7.2).

In the prototype architecture, it is not assumed that primary input patterns are already in space-time form (volleys of spikes). Rather they may be in a purely spatial form—for example, a grayscale image is entirely spatial. Consequently, at the system input, *Input Encode*, converts spatial input patterns to space-time form, i.e., to spike volleys. If one were designing a TNN where the input patterns do not need encoding, say the inputs come directly from another TNN, then the Input Encoding step is not needed (refer to Sections 2.6.3 and 2.6.4).

After the input data is encoded to volleys, the TNN organizes the volleys into clusters via a feature space extraction and reduction process as volleys pass through layers of columns from input to output. A large feature space at the network inputs is reduced to a much smaller set of features at the network's outputs. In other words, the input data patterns are organized into clusters that are small enough to characterize input similarities very compactly. There may be only a single bundle at the output (although reduction to a single bundle is not a strict requirement). The actual characteristics of the clusters (cluster count, especially) are application-dependent.

All the information flow within the TNN is in space-time form. To make the information content more meaningful both for TNN development and for final applications, volley decoders may be attached to any bundle in the network. These decoders provide clustering accuracy metrics as desired.

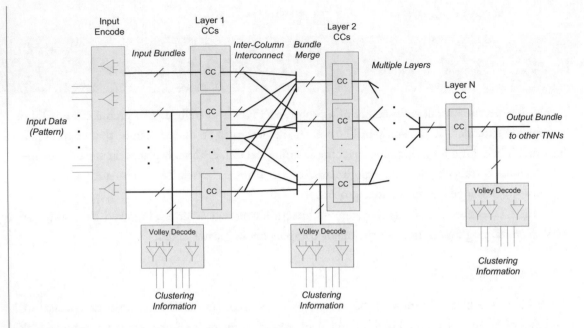

Figure 7.1: Prototype TNN constructed as a series of layers. Each layer consists of a number of parallel computational columns (CCs).

The inputs and outputs of all the TNN components are *bundles* of *lines*. The basic unit of information transfer is a spike volley, which is transmitted on a bundle. As a matter of notation, a bundle will typically be represented as a single line with a slash (although slashes are occasionally omitted to reduce clutter). When it is relevant, the slash will be labeled with the number of lines in the bundle. There is at most one spike per line per volley. Beginning at the primary inputs, volleys pass through the network as a wave of spikes, which are transformed through layered sequences of excitation and inhibition.

A CC is composed of an *Excitatory Column* (EC) and two Inhibitory Columns (ICs); see Figure 7.2. The IC at the input performs feedforward inhibition (FFI), and the IC at the output performs lateral inhibition (LI). The EC is a set of parallel excitatory neurons that perform the column's active computation. Beginning with a spike volley at their inputs, the neurons in an EC generate an output volley according to the functions implied by trained weights, connectivity, and neuron parameters.

Inhibition in an IC operates at a higher level of abstraction than the EC's neurons; i.e., there are no individual inhibitory neurons. The IC as a basic unit analyzes input volley characteristics such as total number and/or timing of spikes, and then modifies the volley passed to the output by removing certain spikes, possibly all of them, while leaving the timing of remaining spikes unchanged.

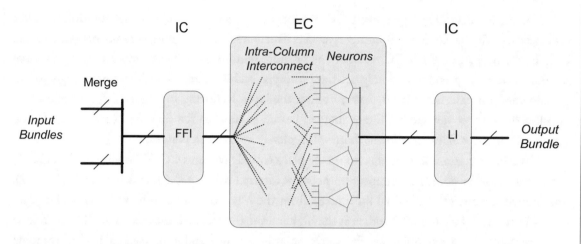

Figure 7.2: A CC is composed of an EC and two ICs. An EC consists of spiking excitatory neurons. The ICs operate in bulk and filter volleys by removing spikes as they pass through from input to output.

The clustering function was described for single excitatory neurons in Chapter 6. A single excitatory neuron forms a cluster within a feature space via unsupervised training. For an EC, which contains a parallel set of excitatory neurons, the function is extended.

With unsupervised training, the training volleys are uniformly applied to all the excitatory neurons in the EC, subject to the connections present in the intra-column interconnection network. An EC can identify multiple clusters, depending on the spiking output pattern. In effect, the EC's neurons implicitly determine the clusters and the output feature space, based on the training volleys and the EC's structure. In evaluation mode, input volleys are applied to the trained EC and the internal neurons produce an output volley indicating its cluster.

7.2   INTERCONNECTION STRUCTURE

There are two levels of interconnects in the prototype architecture. First, there are bundle-level *inter-column* connections among CCs (Figure 7.1); second, there are intra-column connections from the FFI column outputs to the ECs within the same column (Figure 7.2). Determining the TNNs interconnection structure at both levels is one of the more important aspects of the overall TNN design process.

All inter-column communication is at the level of bundles. The input decoding process yields a set of bundles, typically divided according to application dependent receptive fields (RFs), which are essentially subsets of the entire input pattern. The bundles are distributed to the first layer of CCs, one bundle per CC.

Then, in successive layers, each CC produces an output bundle, and the multiple bundles produced by all the CCs in an entire layer are distributed via the inter-column network to the CCs in the next layer. Each CC receives one or more bundles (usually several) according to some regular pattern or pseudo-randomly. In the prototype architecture, a combination of regular and pseudo-random patterns is used, depending on the layer. Referring to Figure 7.2, multiple input bundles directed at the same CC are merged before passing to the first feedforward IC. In the figure, two input volleys are shown; there will typically be more in practice.

The intra-column interconnections take place entirely within a CC. Following the initial FFI, there is a point-to-point interconnection network internal to the EC (dotted lines in Figure 7.2). This network connects individual lines taken from the merged input bundle to individual neuron inputs in the EC. Just as with bundles at the higher level, the selected lines for each CC may follow some regular pattern: a template or tiles, or they may be pseudo-randomly selected. In the prototype architecture, pseudo-random patterns are used.

7.3 INPUT ENCODING

Encoding input information as a spike volley was touched on in Section 2.6.3. Because input information can take practically any form, either numerical or symbolic, there is no single encoding method that can be universally applied. Furthermore, the implemented neuron function is co-designed along with the information coding method, so the encoding method must be carefully thought-out as part of the overall design process.

The prototype architecture implements an encoding method that covers a broad class of input information. In particular, the class includes input information that is presented as vectors of ordered numerical values. These might be drawn from sensors over a two- or three-dimensional surface, for example. They could be a set of voltage levels, sound levels, or intensity levels from touch. Or they could be a stock's price versus time—a one-dimensional input.

A grayscale image is an easily visualized example where information is encoded as an array of ordered values (e.g., from 0–255 for 8-bit grayscale). Consequently, a grayscale image is essentially a surrogate for any of the general class of numerical patterns. For example, outputs from an array of temperature sensors could be visually represented as a grayscale array.

An obvious encoding of numerically ordered, spatial input patterns converts them directly into volleys as done in Section 6.2.1. For example, if there are a total of m input elements (e.g., pixels in a grayscale image), then the value of each of the elements is translated into a spike time. Say the m input values are represented as the image vector $I = <i_1, i_2, \ldots i_m>$, $0 \leq i_k \leq I_{max}$. Then a normalized input volley $X = <x_1, x_2, \ldots x_m>$ is formed directly from input I, possibly with scaling (e.g., V_{max}/I_{max}). That is, when expressed as a value volley, $X_v = <x_1, x_2, \ldots x_m>_v$; $x_k = i_k * (V_{max}/I_{max})$.

Although very straightforward, there are two significant problems with this translation method. As has been observed several times, temporal coding is based on the principle that the first spike in a volley indicates the most important feature, and later spikes have decreasing importance depending on when they occur. However, in a purely spatial pattern, as is the case with grayscale images, the pixel that maps to a spike with the highest value is not necessarily the most important; in fact, all the values may have similar importance when trying to discern similar images. For example, an arbitrary black pixel in a grayscale pattern may contain the same useful clustering information as an arbitrary white pixel. Another example is stock price data; a high price doesn't necessarily provide more useful information than a low price.

A second problem is that the encodings for distinctly dissimilar patterns may have ordering relationships that tend to map them into the same cluster. The volley ordering relationship of interest is described below.

A normalized value volley X^1 *covers* normalized volley X^2, written $X^1 \geq X^2$ if $x_i^1 \geq x_i^2$ for all i. If two volleys do not have a covering relationship, they are said to be *unordered*. The covering relation for a time volley is similarly defined, given the simple inverse linear transformation between value and time volleys.

Different clusters tend to be more easily distinguishable and readily separated if they are identified by unordered volleys. An intuitive argument is that if two patterns should be in different clusters, then each pattern should have at least one feature that is stronger than in the other pattern. If this is not the case, so that the coding for patterns belonging to two different clusters are ordered, then all the features of the covered cluster must also be features of the covering cluster. Furthermore, all the features of the covering cluster are at least as "strong" as for the covered cluster. This suggests potential overlap between the clusters that may make it difficult to separate them.

To solve these problems, one can take an approach that is akin to on-center and off-center coding in the retina (Section 3.5 and [102, 65]). In effect, a negative of the input image is appended to the original (positive) input image, resulting in an image that doubles the volley size. This is illustrated in Figure 7.3.

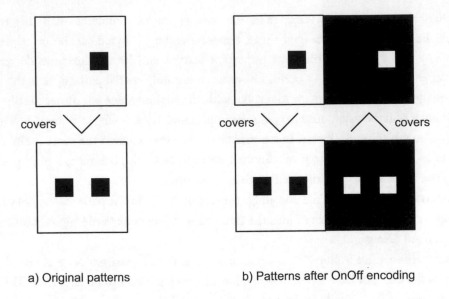

a) Original patterns   b) Patterns after OnOff encoding

Figure 7.3: In the original patterns (a), the volley coding of the top pattern covers the bottom one. Then with OnOff coding (b), the negative of the bottom pattern necessarily covers the top. Hence, it is impossible for there to be any ordering among OnOff encoded patterns.

More precisely, if the original input volley is: $X = <x_1, x_2, ... x_m>$, then the OnOff encoded volley is $X = <x_1, x_2, ... x_m, x_m+1, x_m+2... x_{2m} >$, where $x_{m+k} = I_{max} - x_k$, where $I_{max} = V_{max}$ if X is a value volley, and $I_{max} = T_{max}$ if X is a time volley.

Example

Given a spatial input pattern with values $I_v = <2, 16, 6, 0>_v$, where $I_{max} = 16$, following is the formation of the input volley X_v. First, for the sake of discussion, assume $V_{max} = 8$. Then all the values in I are scaled by $8/16 = .5$, so the scaled value volley for input I is: $X_v = < 1, 8, 3, 0>_v$. The negative volley for X_v is: $< 7, 0, 5, 8>_v$. The scaled input volley and its negative are then concatenated to form the encoding input volley $X_v = < 1, 8, 3, 0, 7, 0, 5, 8 >_v$.   □

After this encoding is performed on all original grayscale patterns, there is no temporal ordering among pairs of patterns, and each element in the original pattern contributes equally to the new (2x larger) pattern. That is, a high value pixel is always paired with a comparably low value pixel, and vice versa.

7.4   EXCITATORY COLUMN OPERATION

In normal operation, input volleys are applied to an excitatory column to train them by establishing synaptic weights, and then the same input volleys may be applied to the trained column to identify the clusters to which they belong. It is most convenient to discuss these two modes: training and evaluation, in reverse order.

7.4.1   EVALUATION

Assume that prior to evaluation, the excitatory neuron weights, parameters, and interconnection structure have been established through a training process.

In evaluation mode, each of the excitatory neurons in an EC performs its evaluation independently of the others. A single bundle is directed to a column, and internal to the EC a subset of the lines in the bundle are directed to the individual neurons in the column. Consequently, depending on the intra-column interconnections each neuron sees some subset of the spikes in an input volley.

Evaluation relies on temporal invariance (Section 2.4). Assume the evaluation process begins with unnormalized input volleys where spike times are global times. Then, for a given neuron, the evaluation process first normalizes its input volley (to yield local times), performs evaluation, and then de-normalizes the output back to global time. The de-normalized outputs of all the neurons in the EC are combined to form the EC's output volley in unnormalized form expressed in global time.

Assume a time volley, $X_t = <t_1, t_2, ... t_m>_t$ is input to a neuron. The spike with the minimum time is first determined: t_{min}. Then t_{min} is subtracted from each of the spike times in the volley to normalize it. Consequently, the normalized volley is $X'_t = <t_1 - t_{min}, t_2 - t_{min}, ... t_m - t_{min}>_t$. The excitatory neuron performs its operation, and then the value t_{min} is added back into the output spike time for each neuron. So, if the output of a neuron after evaluation is local actual time t_z, then the final output after de-normalizing is $t_z + t_{min}$. All the individual output spike times are combined to form the output volley.

7.4.2   TRAINING

Generally speaking, training is the process of adapting a TNN to fit the features of the input patterns. Historically, in conventional neural networks, it is the weights that are trained. In the work here, training the weights is important; however, as has noted earlier, weights are not the only part of a neural system that are plastic. Depending on the time scale, virtually every feature of the neocortex appears to be plastic, or adaptable.

Consequently, training as done in the work here is not restricted to synaptic weights. Training also includes a number of neuron parameters as well as the topology of the interconnection network. Synaptic weight training remains very important, however, and forms the inner loop of the larger training regime. Outer loops allow adaptation, or tuning, of other neuron parameters and interconnection structures.

A key feature of the paradigm given here is that weight training is extremely fast. This allows efficient exploration of the larger design space formed by the other, outer loop parameters. Although the characterization of "inner loop" and "outer loop" functions is drawn from a specific design methodology, the outwardly observable effect is a high degree of system-wide plasticity that extends beyond synaptic weights.

A specific inner-loop, outer-loop training schema depends on the application and implementation (the simulator code in our case). Consequently, further discussion of an overall training method is given later in the context of a specific system design (Section 8.6).

In the remainder of this section, attention is restricted to the inner-loop function of synaptic weight training.

7.4.3 UNSUPERVISED SYNAPTIC WEIGHT TRAINING

Unsupervised weight training is performed at the EC level. Consequently, one applies all training patterns to all the excitatory neurons in the column. If all the neurons in an EC are identical, including the input interconnection structure, then they will all wind up with identically trained synapses and all the neurons will respond identically to any applied input volley. This is obviously not desirable if one wants to separate clusters.

It follows that a key architectural aspect of unsupervised training at the column level is that there must be some variation within the column. For example, there could be variation among individual neuron parameters that affect the training of weights and/or there may be some variation the column interconnection structure. In the prototype TNN, rather than using complete point-to-point connections from all input lines to all neurons in the EC, pseudo-randomly selected subsets are used.

For the excitatory neuron model used here, all the simple synapses forming a compound synapse are combined into a higher-level model—the compound responses as represented by the E_i functions (Section 6.6). Synaptic training then becomes a process of simply finding mean spike times. Consequently, the method for unsupervised training takes the average (arithmetic mean) value for spikes belonging to each of the lines over the entire training volley set, and then scales them so they fit within the maximum weight range defined by W_{max}. Scaling using the ratio $\gamma = T_{max}/W_{max}$ is one approach, and the one used here, although one could scale in other ways, as long as the final weights are between 0 and W_{max}.

Denote the training volley set as R; then there is a total of $|R|$ volleys in the training set. For a given input line i (which feeds a compound synapse), let t_i be the spike time within a given normalized training volley. The weight associated with this spike is $(T_{max}-t_i)/\gamma$. The overall weight for the synapse associated with line i is then the average of the associated weights, i.e., $w_i = \sum (T_{max} - t_i) / (\gamma \mid R \mid)$, where the summation is over all the volleys in the training set R. Equivalently for value volleys, the incremental weight contribution is $(V_{max} - v_i) / \alpha\gamma$, and the average synaptic weight for input line i is $w_i = \sum (V_{max} - v_i) / \alpha\gamma \mid R \mid$.

Because training amounts to taking the arithmetic mean of the spike times, each training pattern needs to be applied only once, and the averaging process is computationally very simple. Consequently, in a simulator, weight training is extremely fast.

Example

Assume a neuron with four compound synapses. $T_{max} = 16$ and $\varepsilon = .1$. $W_{max} = 8$, so $\gamma = T_{max}/W_{max} = 2$. The training set is $\langle 2,0,9,4 \rangle_t$, $\langle 0,1,8,6 \rangle_t$, $\langle 3,0,7,4 \rangle_t$, $\langle 0,2,9,2 \rangle_t$. Then, there are four weights to compute, one for each of the compound synapses.

$w_i = \sum (T_{max} - t_1) / (\gamma \mid R \mid)$, summed over all training volleys. This is $w_1 = (14 + 16 + 13 + 16)/(2*4) \cong 7.4$ (round to .1). Similarly, $w_2 = (16 + 15 + 16 + 14)/8 \cong 7.6$; $w_3 = (7 + 8 + 9 + 7)/8 \cong 3.9$; $w_3 = (12 + 10 + 12 + 14)/8 \cong 6.0$. So, the weight vector for the neuron is $\langle 7.4, 7.6, 3.9, 6.0 \rangle$. This weight vector effectively implies the cluster center. Expressed as a time volley the implied center is: $\langle 1.2, 0.8, 8.2, 4.0 \rangle_t$.

7.4.4   SUPERVISED WEIGHT TRAINING

Although the prototype TNN uses unsupervised weight training, one can also employ a supervised version. First observe that from the perspective of each individual excitatory neuron, all training is always unsupervised. The neuron learns from everything it observes at its inputs. Consequently, supervised training is implemented merely by restricting what the neuron observes. That is, during supervised training, only a selected proper subset of input training patterns are applied to a given individual neuron.

Thus, every training pattern for which a given excitatory neuron should emit a spike belongs to that neuron's supervised training set (and all other training patterns do not). The sets of training patterns may be determined by input meta-information, e.g., labels.

For supervised training, then, a specific subset of the training volleys is selected for each individual excitatory neuron, depending on the problem at hand. For example, we might have a training set containing labeled input patterns. L different labels for the training set identify L different classes. Then, we can build a classifier network with ECs consisting of L excitatory neurons, $E_1 ... E_L$, one for each of the labels. Neuron E_j is trained by applying only members of the training with label $L = j$. Let R_j, $j = 1..L$, be the set of training volleys for which neuron E_j should emit an output spike. The averaging equations for determining weights are then similar to those for unsupervised training, except that averaging is over the training subset R_j. Then, $w_i = \Sigma \, (T_{max} - t_i) \, / \, \gamma \mid R_j \mid$, where the summation is over all the volleys in R_j.

By doing this, each neuron will be more highly attuned to the patterns for which it should emit an output spike. When an evaluation pattern is applied, then the neuron(s) trained with similar patterns will spike earlier than neuron(s) trained with dissimilar patterns (if they spike at all).

7.5   INHIBITION

Thus far, inhibition has gotten short shrift. It has taken this long to get to an expanded discussion of inhibition because an architectural context is required for inhibition to make sense. In practice, inhibition is always combined with excitation. Inhibition appears as a bulk process where many inhibitory interneurons act in concert to implement a function that is column-wide. That is, model inhibition works at the column level, not the individual neuron level.

As described above, each excitatory neuron in an EC acts independently of the others. The only things the individual neurons share are inputs. In contrast, inhibition performs *collective* operations on all the spikes in a volley. It is loosely analogous to pooling in conventional deep learning networks [86]. Overall, the ECs generate spike volleys from independently acting excitatory neurons, and inhibition makes modifications to the volley based on collective timing relationships across the spike volley. The output volley of an IC is determined by timing relationships among spikes belonging to its input volley. In the prototype TNN, the inhibition implements filtering operations that remove selected spikes from a volley (possibly all the spikes), based on the volley's initial spike pattern.

Recall that in biological neurons, inhibition has the opposite effect of excitation. That is, an incoming inhibitory spike reduces a neuron's membrane potential, thereby decreasing the likelihood of an output spike. Before going into detail, following are some general effects of inhibition.

One obvious first order effect of inhibition is that it reduces the total number of spikes that a CC produces and are propagated to other neurons. Doing so may save energy (in a direct physical implementation) and/or time (in a simulation model). In this early stage of research, however, emphasis is more on functionality rather than efficiency.

Inhibition may also eliminate spikes in a volley that carry relatively little information, thereby enhancing the effects of the more information-rich spikes that remain. There are at least two situations where this can occur.

1. If spikes in a volley are spread over time, the spikes occurring later in time offer less useful information than the earlier-occurring spikes. Hence, one might use inhibition to filter out some of the later occurring spikes. By eliminating these spikes, clusters become more focused, in a sense.

Input patterns are forced into a smaller set of clusters when later spikes are removed.

2. If many spikes occur nearly simultaneously at the leading edge of a volley, then these spikes also carry relatively little information. In the extreme case, if all the neurons in a column produce a spike at the same time, there is essentially no more information than if none of them spikes; all of the features are equally strong. In these cases, eliminating all the simultaneously occurring spikes may improve functionality.

A large number of low-information spikes may drown out smaller numbers of high-information spikes, so removing them allows the high information spikes greater influence on the output.

One *could* model inhibition with individual inhibitory neurons just as excitatory neurons are (some theoreticians and modelers do this). However, in the computation model described here, inhibition is not implemented with individual inhibitory neurons and responses. Rather, inhibition is implemented as a bulk process at the column level. Inhibition acts on a full column of excitatory neurons, and it operates on entire volleys rather than individual spikes. Karnani et al. [45] call this bulk operation a "blanket" of inhibition—a very descriptive term.

Generally speaking, there are three types of inhibition: feedback, lateral, and feedforward. At a primitive level, these were illustrated much earlier in Figure 3.6.

7.5.1   FEEDBACK INHIBITION

In the context of the TNNs in this book, feedback inhibition can be dispensed with immediately. For completeness, feedback inhibition at the column level is shown in Figure 7.4. With feedback inhibition, the output spikes of an EC (right side of figure) are fed back to their respective upstream neurons in order to inhibit them from any further spiking. That is, an excitatory neuron emits an output spike that is fed back to inhibit *itself* from spiking again. In the biological neocortex, such feedback may be important for dividing spike trains into distinct volleys, as well as saving energy. However, we use a model where division into volleys is incorporated directly in the model.

Consequently, in the TNN model, such feedback inhibition would have no effect on an excitatory neuron because in a TNN, a given excitatory neuron produces at most one spike per volley. Hence, its spike output cannot possibly influence any further spikes from the same output.

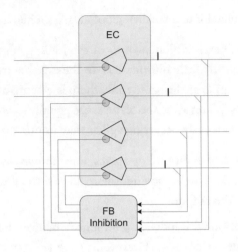

Figure 7.4: Feedback Inhibition (FBI).

7.5.2   LATERAL INHIBITION

Lateral inhibition (Figure 7.5a) uses the initial spikes from some subset of excitatory neurons in an EC to inhibit spiking of *other* excitatory neurons belonging to that same EC. The *effect* of lateral inhibition (LI) is to allow early spikes in a volley to propagate downstream to other columns, while eliminating later spikes from the same volley.

This is a function that is commonly implemented in proposed TNNs. Lateral inhibition eliminates all but the first k output spikes of a volley produced at the EC's outputs. This is typically referred to as k-winner-take-all (k-WTA), because only the first k spikes, the "winners", are allowed to proceed to the next column.

An extended form of lateral inhibition adds a temporal parameter, to arrive at tk-WTA. The temporal parameter t_L specifies a time interval measured with respect to the first spike in a volley. Only spikes occurring within time t_L of the first spike are included in the LI's output volley. Then, a second parameter, k_L, further restricts the output to the first k_L spikes.

For a normalized volley input $<t_1, t_2, t_3, ... t_{n-1}, t_n>_t$,

$$LI\ (<t_1, t_2, t_3, ... t_{n-1}, t_n>_t\ ,\ t_L, k_L) = <t_1', t_2', t_3', ... t_{n-1}', t_n'>_t\ ,$$

where $t_i' = t_i$ if $t_i \le t_L$ and t_i is one of the first k_L spikes, otherwise $t_i' = $ "-".

The generalization to unnormalized volleys is straightforward.

Example

$$LI\ (<4, -\ , 6\ , -\ , 0\ , -\ , 1\ , 8\ , 5 >_t, 5, 3) = <4, -, -\ , -\ , 0\ , -\ , 1\ , -\ , - >_t\ \square$$

Lateral inhibition discards information—it removes spikes. However, the discarded information has less information content than the spikes that are kept. This reduction in spikes typically results in a reduction in the number of clusters.

Finally, note that feedback and lateral inhibition are distinguished only for modeling purposes. In the biological setting, they aren't physically separate. All of the neurons inhibit all of the other neurons (including themselves). In essence, feedback inhibition and lateral inhibition become one-and-the-same.

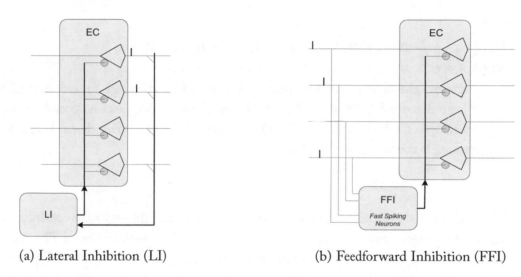

(a) Lateral Inhibition (LI)   (b) Feedforward Inhibition (FFI)

Figure 7.5: Two forms of inhibition used in the TNN model under development.

7.5.3 FEEDFORWARD INHIBITION

A feedforward inhibition block (Figure 7.5b) takes the same inputs as the associated EC in its CC. It then inhibits some (or all) of the neurons in the associated downstream EC from spiking, depending on the intensity of the input pattern that the inhibitory neurons receive. Depending on the number of spikes invoking inhibition, some (possibly all) of the EC's output spikes are eliminated.

The important *effect* of feedforward inhibition (FFI) is to eliminate the downstream propagation of all spikes in a volley if the volley contains a relatively large number of spikes that occur very early and nearly simultaneously. Biologically, the large number of early spikes are propagated through fast spiking feedforward inhibition (see Section 3.2.4) which chokes off the downstream EC before any of its neurons can produce an output spike (refer to Figure 7.5b).

FFI is modeled with two parameters, t_F and k_F. If more than k_F spikes in a volley occur within time t_F of the first spike, then all spikes in the volley are eliminated ("*nullified*"). Otherwise, all the input spikes are allowed to pass to the output, unchanged.

For a normalized volley input $<t_1, t_2, t_3, ... t_{n-1}, t_n>_t$,

$$FFI (<t_1, t_2, t_3, ... t_{n-1}, t_n>_t , t_F, k_F) = <t_1', t_2', t_3', ... t_{n-1}', t_n'>_t,$$

$$\text{let } L = \{t_i \mid t_i \leq t_F \},$$

$$\text{then if } |L| \leq k_F , t_i' = t_i \text{ for all } i;$$

$$\text{else } t_i' = \text{"-" for all } i.$$

The generalization to unnormalized volleys is straightforward.

Examples

$$FFI (<4, -, 6 , - , 0 , - , 1 , 8 , 5 >_t, 2 , 2) = <4, -, 6 , - , 0 , - , 1 , 8 , 5 >_t$$

$$FFI (<4, -, 6 , - , 0 , - , 1 , 8 , 5 >_t, 2 , 1) = < -, -, - , - , - , - , - , - , - >_t \qquad \square$$

Training determines the exact parameter values t_L, k_L, t_F, and k_F. Although operation of inhibition has just been described, training of inhibition has not. The training process depends on volley decoding and analysis methods, which are described in the next section. With that as background, inhibition training is covered in Section 7.7.

Causality

The FFI model as just described is not causal. However, it is described this way to be consistent with the simulator that implements the model. An equivalent causal implementation of FFI and EC that reflects the biology more closely is illustrated in Figure 7.6 (FFI′ and EC′). Here, FFI′ counts the number of input spikes that are less than or equal to the minimum input spike time plus the parameter t_F. If this number exceeds parameter k_F, then the single output is $t_{min} + t_F$. Otherwise the output is nullified. Then EC′ nullifies any output spike at $t_i < t_{min} + t_F$. As long as t_F is less than the minimum latency through EC′, then the net overall effect will be the same as in the non-causal FFI model described above. For simulations performed later in this book using the non-causal FFI, this condition always holds.

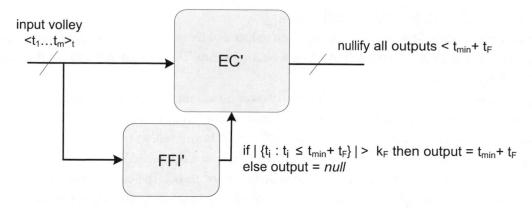

Figure 7.6: Causal FFI model.

7.6   VOLLEY DECODING AND ANALYSIS

A volley in a TNN is a space-time encoding of cluster information (1) using a feature set implied by TNN structure and training, and (2) with feature values encoded in space-time form. For these two reasons, it is extremely difficult for us to extract meaningful information directly from the outputs of a TNN or from internal bundles. If the TNN directly feeds another TNN, then this does not pose a problem; we do not have to be directly involved or understand the cluster encoding. However, if we want to use the output information, say as part of machine learning application, then we must make the TNN space-time output more comprehensible by *decoding* it.

A *volley decoder* takes a space-time volley as input and produces a useful characterization of the information being conveyed by that volley. What is "useful" depends on the application, of course. For simple clustering as envisioned here, the information consists of output spike patterns that indicate, or identify, a particular cluster.

As illustrated for a generic TNN architecture (Figure 7.1) volley decoding serves three purposes. The first is to provide performance feedback to the designer as TNN development progresses; this is illustrated in the figure with decoders attached to two of the TNN's internal bundles. Any number of internal bundles can be decoded concurrently, each with its own decoder. The second is to provide optimal parameters to the inhibitory training process during simulation; here, decoding is internal to the CC implementation and is not shown in the figure. However, this will be discussed in some detail, in Section 7.7. The third is as part of the last stage of a machine learning classifier. That is, if one were to use the TNN's clustering function as part of a machine learning classifier, then a *classifying decoder* is attached to the TNN's primary output.

7.6.1   TEMPORAL FLATTENING

The first step of decoding, as proposed here, is to reduce and temporally flatten a volley's space-time information into a purely spatial form—into a binary pattern. These binary patterns are then used for identifying clusters.

Because the earlier spikes of a volley carry most (or all) meaningful information, later spikes can be removed without diminishing the volley's information in any significant way. After removing slower spikes, only a small number of spikes will remain in the volley (1 to 4, say). Furthermore, these earlier spikes typically occur close together in time, so a further reduction is to assume that they occur at exactly the same time (i.e., temporally flatten them). The combination of removing slow spikes and assigning identical times to the remaining spikes reduces a volley to a simple, sparse binary vector. This process is illustrated in more detail in Figure 7.7.

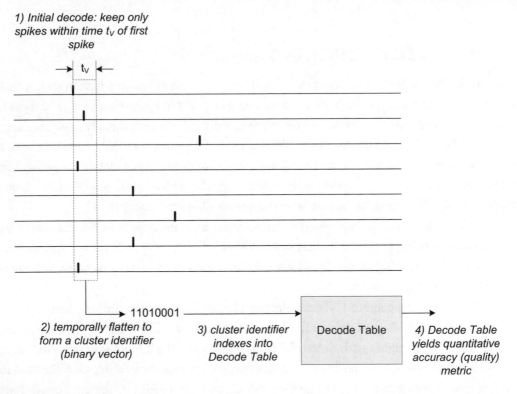

Figure 7.7: With decoding, volleys are first reduced to only the earliest spikes, which are then flattened into a binary vector. The binary vector is a cluster identifier. This vector accesses a Decode Table which yields a performance metric, e.g., clustering accuracy.

In Figure 7.7, the decoding step is accomplished by first defining a time parameter t_V. Any spikes occurring within t_V of the very first spike in the volley are kept; all the others are eliminated.

The remaining spikes are then reduced to a binary vector, thereby discarding the actual times at which they occur. The resulting decoded binary patterns are purely spatial. These binary patterns are referred to as *cluster identifiers* because all the TNN input patterns that result in this particular binary vector are considered to belong to the same cluster.

It is important to note that this reduction to a cluster identifier is *lossy* both because some spikes in the decoded volley are discarded and because the remaining spikes have all temporal differentiation eliminated. Hence, later, when we deal with various performance metrics based on this decoding, it should be inferred that all the metrics are *lower bounds* on the information quality that flows along the bundle.

7.6.2   DECODING TO ESTIMATE CLUSTERING QUALITY

During research and development, we are most interested in decoding volleys so we can estimate the quality of the clustering that has been performed up to some point (i.e., a bundle) in a network. For that reason, we would like quality metrics—i.e., to measure the quality of input pattern *similarity* that the clustering function has identified. Throughout this section, when refer to an "input" or "input pattern", we are referring to the TNN's primary input, not an EC's local input; it is the TNN's primary input that we are attempting to place into clusters.

A simple way of assessing quality is to use some method of subjectively evaluating similarity—by visually inspecting and judging similarity on a case-by-case basis. When we are dealing with grayscale images as input patterns this human-in-the-loop subjective method has proven to be a very useful technique for acquiring feedback and insight. As an example, this method is used later in Chapter 9, when a specific TNN design is discussed in detail.

As far as quantitative metrics, useful metrics can be defined by using meta-data (e.g., input labels) as a standard for comparison. Then, clustering accuracy can be quantitively measured against that standard.

For example, consider the clustering of Arabic numerals as summarized in Figure 2.6 and associated text. The input is a spike-encoded grayscale pattern for a handwritten numeral 1-9, and the output is a small binary code word that identifies a particular cluster to which the input pattern belongs. Ideally, we would like a small number of clusters (at least 10 if we are identifying the numerals), but the number of clusters could be more than 10, say as high as 20 or 30 for this example.

What we want is for input patterns to be clustered together according to similarity. The meta-data labels provide an external *opinion*, so-to-speak, regarding image similarity. If two inputs have the same labels, then they may be considered *similar*. So, if we accept the meta-data (labels) as a definition of similarity, our ideal is for each of the clusters to contain input patterns all having the same label. In practice, of course, this ideal will be very rarely achieved.

Failing the ideal, we would like some measure of how far we are from achieving the ideal. To do this, observe that network training (clustering) yields a mapping from network inputs to cluster identifiers. Within a cluster, the largest group of mutually similar input patterns all have the same label (by definition of similarity). Any pattern with a different label becomes an outlier with respect to the largest group. Hence, we consider the label for members of the largest group to be *correct* in a sense, and all the outliers as being *incorrect* with respect to the cluster at hand. The *clustering accuracy* for the given cluster is then the ratio of correct input patterns to all input patterns (almost). To be more precise, we should say "all input patterns that yield a non-null cluster identifier," because, following inhibition, some spike volleys traversing a bundle may have no spikes at all, i.e., they are null.

If a volley is null, it is assumed to carry no useful clustering information. Consequently, *coverage* is defined to be the fraction of all volleys that are information-carrying non-null volleys. In essence, a null volley conveys a "no cluster" message, indicating an input pattern that falls outside all the clusters. To reiterate: *coverage* is the fraction of volleys that are non-null (i.e., that contain at least one spike).

Regarding implementation, clustering accuracy and coverage may be computed via a *Decode Table*. The details are best given by a small-scale example.

Example

Figure 7.8 contains the Decode Table for spike volleys containing six lines. In this example, input patterns belong to one of four classes, labeled A, B, C, and D. Each row in the decode table is associated with a specific cluster identifier.

The four table columns labeled A–D contain the numbers of times the given label occurs in conjunction with the corresponding 6-bit cluster identifier. The *total* column is the number of input volleys that decode to the given cluster identifier. The *maximum* column is the maximum count over all the label counts. The particular label that has this maximum is defined to be *correct*. Then the maximum divided by the total is the *clustering accuracy with respect to the input labels*. Although we will often shorten it to "cluster accuracy," it is important to note that this accuracy is only defined with respect to a given set of meta-data labels.

So, in the example, the spike pattern 000110 occurs for a total of 686 out of 2,750 input patterns that yield a non-null volley. Of these, 49 are labeled A, 116 are labeled B, 192 are C, and 329 are D. Because the label D is the most frequent, the correct similarity group is defined to be D (the entry in the *similarity group* column).

Overall, there are a total of 2,750 input patterns that yield a non-null output, and the total of the maxima is 1,785. Hence, the overall clustering accuracy is 1,785/2,750 = .65 . If in this example, we assume the total number of input patterns is 3,000 (so 3,000 - 2,750 = 250 result in null volleys). Then the coverage is 2,750/3,000 = .92 □

cluster identifier	label counts				total	maximum	similarity group
	A	B	C	D			
000011	11	34	58	0	103	58	C
000110	49	116	192	329	686	329	D
001100	41	71	426	11	549	426	C
001001	4	2	35	0	41	35	C
001010	0	6	160	7	173	160	C
011000	15	397	36	2	450	397	B
010100	0	24	2	0	26	24	B
010010	56	0	6	2	64	56	A
010001	110	0	95	45	250	110	A
110000	0	82	65	5	152	82	B
100100	1	0	1	0	2	1	C
101000	6	3	6	0	15	6	C
100001	101	58	18	62	239	101	A
					2750	1785	
				accuracy =		0.65	

Figure 7.8: Decode Table example.

As an important part of the process, the time parameter t_V is tuned to maximize the accuracy metric. This tuning is done in practice by constructing a Decode Table for a range of t_V values, determining the resulting accuracy of each, and selecting the t_V that yields the best accuracy. In simulation, this can be done very quickly.

7.6.3 DECODING FOR CLASSIFICATION

Eventually, we may want to construct a classifier built around a clustering TNN. The detailed implementation of a classifier is not pursued in this book. Following is a summary of the way the clustering TNN can be converted, so to speak, into a classifier.

Briefly, we first use unsupervised training to identify clusters in the normal way. Then, given the trained TNN, we apply *new* evaluation patterns to the clustering TNN's inputs and the TNN will yield a cluster identifier decoded as described above.

A major difference between application of the Decode Table for measuring cluster accuracy and for performing classification is that with classification, one of the never-before-seen evaluation inputs may decode to a binary vector that does not match any of the cluster identifiers present in the Decode Table.

This means that one must consider interpolation methods between the given decoded volley and non-empty Decode Table entries. Because decoded volleys are temporally flattened into a binary vector, interpolations based on Hamming distance can be used. There are a number of ways one can interpolate. For example, one can take a majority vote of all the distance 1 table entries, if there are no distance 1 table entries, then try Hamming distance 2, and so forth.

Overall, classification proceeds as follows. After an evaluation input is run through the trained TNN, the output volley is decoded into a binary vector (cluster identifier). If there is an entry in the table that corresponds to the decoded binary vector, one should use the corresponding entry to classify according to the label with the maximum count during training. If there is no table entry for the given decoded volley, then use label frequency data from table entries based on their Hamming distance from the decoded volley. The given input is then classified according to data taken from nearby cluster classification(s).

7.7   TRAINING INHIBITION

For an IC, we have identified four inhibition parameters, t_L and k_L for LI, and t_F and k_F for FFI. As currently modeled, the parameters t_F and k_F are generated specifically for each individual IC, as is the case with weights in the EC. Also as currently modeled, the same parameters t_L and k_L are used for all ICs in a layer; however, this could change as the model is developed further.

Because ICs are modeled at a higher level than individual neurons, the training process is also performed at a higher level.

7.7.1   FFI: ESTABLISHING t_F AND k_F

First, consider the qualitative effects of t_F and k_F. Lower values of t_F and k_F tend to nullify more volleys. A lower t_F reduces the numbers of spikes that are compared against k_F, and a lower k_F increases the chances that a volley will be nullified. Consequently, lower values of t_F and k_F tend to reduce coverage.

Meanwhile, accuracy for the non-nullified volleys tends to increase. After nullified volleys have been removed, the remaining volleys tend to have higher information content because they contain fewer spikes, and the spikes that remain are all fast spikes. Therefore, they tend to yield better decode accuracies.

As a consequence of the above observation, FFI can be employed to improve the overall quality of information in a volley as it passes through inhibition. That is, FFI essentially filters out lower information volleys by nullifying them, leaving only higher information volleys. Training then becomes a process of identifying t_F and k_F values that in some way determine an optimal decode accuracy/coverage pair.

The two FFI inhibition parameters define a two-dimensional accuracy/coverage tradeoff space, as illustrated in Figure 7.9. The first step of the IC training process consists of generating a set of accuracy/coverage data points, one for each pair of t_F and k_F values, where t_F and k_F vary over a designer-specified range. Currently, this is done by exhaustively evaluating a wide range of parameter pairs, and establishing the accuracy/coverage data point for each. The computation time for establishing each data point is very short, so exhaustively covering a reasonable space is not burdensome. Furthermore, it is likely that a more efficient, non-exhaustive approach can be developed. As far as the current research, however, doing so does not have a very high priority, given that the exhaustive method is fast enough.

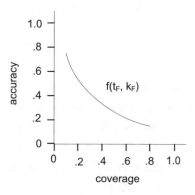

Figure 7.9: Generic accuracy/coverage tradeoff curve, determined by the space of t_F, k_F values.

After the coverage and accuracy for a set of $<t_F,k_F>$ pairs are determined, the set of data points are then greatly reduced by eliminating all but the Pareto-optimal points. A Pareto-optimal curve is generically illustrated in Figure 7.9.

Given the Pareto-optimal data points, the next step in the process is to select one. There are a number of approaches that could be used. For the prototype TNN, one in particular is being used.

Consider FFI training being applied to the set of volleys that pass from one layer to the next. For the set of volleys, we define an overall *iso-accuracy value*. The iso-accuracy point for a given IC is the pair $<t_F,k_F>$ that achieves the highest coverage while providing *at least* the desired iso-accuracy value. The t_F and k_F values that produce this iso-accurate data point then become the parameters for the FFI in question, thereby training the FFI. What makes this parameter "iso" is that this same accuracy point is used for all the ICs in the layer under development.

7.7.2   LI: ESTABLISHING t_L AND k_L

In the current design, lateral inhibition is used for forcing relatively sparse CC output volleys, thereby reducing the number of clusters. This has the desirable effect of focusing a large number of input patterns into a much smaller set of clusters.

The current method for establishing the parameters is done in two steps. First, a value of k_L is assigned. This can be done manually in an outer loop of the optimization process. The value of k_L directly determines the sparsity of output volleys, and indirectly determines the number of clusters.

Then, after k_L is established, as part of inner loop automatic optimizations, a range of t_L values are analyzed by the volley decoding process outlined above, and the one yielding the best clustering accuracy is chosen.

Note that LI never leads to nullification of a volley because for any non-null input volley there is always at least one spike at $t = 0$ (after normalization). As with FFI, the computation for a given t_L is relatively fast so a fairly large range of t_L values can be swept in a short time.

7.7.3   EXCITATORY NEURON TRAINING IN THE PRESENCE OF INHIBITION

The training approach discussed thus far overlooks the potential interaction between inhibition and STDP weight training of excitatory neurons. Specifically, with STDP, a weight update should only occur for those neurons that actually produce an output spike. However, some output spikes are removed by inhibition. Consequently, there is a linkage between inhibition and STDP weight updates. This linkage between inhibition and STDP is not present when excitatory neuron training is implemented as proposed here. Rather, weight updates effectively occur as if all neurons always spike.

The effect of updating weights based on the post-inhibited output volleys (after LI) may be accomplished in the model in the following way. First, the training volleys are applied and the weights are (temporarily) fixed at the trained values. Then, the training set is applied again, this time as an evaluation set, with LI present. The spikes that are *not* inhibited by LI form a *training mask*. Finally, training is performed a second time, controlled by the newly formed training mask. That is, a spike in a training volley is used for training only if the corresponding training mask bit is set to one. The weights determined during this final training step become the final weights used for evaluation. In this way, synapse weights are affected only by the output spikes that are not inhibited.

PART III

Extended Design Study: Clustering the MNIST Dataset

CHAPTER 8

Simulator Implementation

The text up to this point has focused on development of a TNN model. This chapter describes a simulator for a prototype TNN architecture. Although the simulator is intended for architecture development, the simulation code could later become the basis for an eventual machine learning implementation.

8.1 SIMULATOR OVERVIEW

A generic block diagram of the simulation system is in Figure 8.1. The overall simulator consists of five types of functional blocks. The left half of the figure shows the input layer, Layer 1, which is specialized to take spatial RFs as input. The right half of the figure is a general layer. Although not shown in the figure, the total number of columns in each layer is fewer than in the preceding layer, by approximately a factor of four in a typical system. Each layer reduces the number of bundles passed to the next layer, until only a single bundle remains.

As an introduction, following are brief summaries of the five block types.

- **OnOff:** This is a general input encoder which translates the inputs coming from purely spatial receptive fields into spike times used by the rest of the system. The individual input signals (e.g., grayscale values) are assumed to be numerically ordered.

- **EC: Excitatory Columns** implement the excitatory column functions as described in Sections 7.4 and 7.4.2. In training mode, EC units take the training input volley sets as input and compute the weights. In evaluation mode, the weights are fixed. Evaluation spike volleys are applied and the EC unit determines the output spike volleys.

- **IC: Inhibitory Columns** perform inhibition on input volleys to produce output volleys. Inhibition may delete some spikes, but otherwise output spikes are unchanged with respect to input spikes.

- **MG: Merge Units** merge multiple volleys into a single volley. This amounts to concatenating the volley vectors and re-assigning spike numbers accordingly.

- **VA: Volley Analyzers** analyze volley sets and generate statistics and plots of the observed behavior. Very importantly, they determine the Pareto-optimal accuracy/coverage data set that is used for inhibition training. The VA units do not modify volleys in any way; they merely tap into a bundle and are an analysis tool for IC training and for guiding the design process.

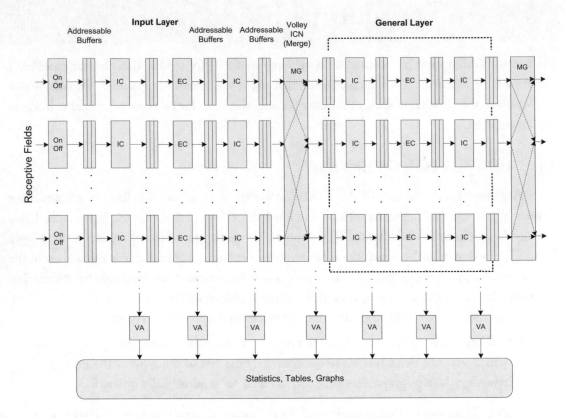

Figure 8.1: Example execution architecture implemented in a software simulation system. Connections between blocks are bundles which carry volleys.

8.2 INTER-UNIT COMMUNICATION

When spike volleys are stored and communicated inside the simulator, the non-null spikes are represented in sparse vector format. For example, the time volley $<4,-,0,-,-,1,-,8>_t$ is represented as a set of pairs: [$<1,4>, <3,0>, <6,1>, <8,8>$], where the first element of a pair indicates the spike position in the volley, and the second indicates the numerical value.

For communication among the functional blocks, volley sets are passed through addressable buffers, where each addressable buffer is associated with a particular volley set. An interconnection structure is implemented via the assignment of buffer addresses. Each EC or IC block is assigned a uniquely identified addressable buffer for its output volley. By using the same unique identifier, any other unit (MG, EC, IC, or VA) can consume the associated volley set as its input. Thus, by assigning volley buffer identifiers, the inter-CC interconnect patterns are established.

In the current simulation model, volley sets are of some fixed length, and addressable buffers are maintained as secondary storage files so they persist from one simulation run to the next, thereby simplifying one-layer-at-a-time TNN development. Consequently, the simulator is structured so that a volley buffer stores a complete volley set before forwarding it to the next unit.

8.3   SIMULATING TIME

Overall, global time is maintained in the simulator as summarized in Section 2.5. Basically, a global time wrapper (Section 2.5) is implemented around each neuron.

At each abstract time step:

1. Spike volleys destined for excitatory neurons within a CC are normalized to form local time frames for each neuron. This is done by finding the minimum spike time in the volley and subtracting it from all the spike times.

2. The CCs operate within a local time frame of reference, due to temporal invariance.

3. The output volleys are placed back into a single global frame of reference by adding the minimum spike time that was subtracted in (1).

An example for evaluation is shown in Figure 8.2.

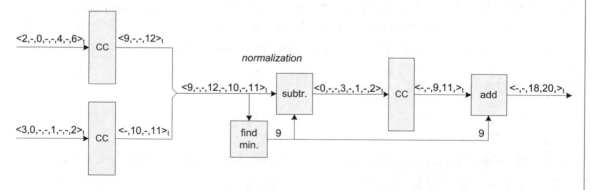

Figure 8.2: Example of global/local time conversion. $T_{max} = 8$. Assume a complete point-to-point interconnect to excitatory neurons in a CC.

If one uses the global time reference $t = 0$ at the inputs of the first layer of the network, as spike volleys pass through multiple succeeding layers, the global actual times will increase monotonically, potentially without bound. For very deep networks, this growth may be undesirable because of the storage and computation resources it may needlessly consume. To mitigate this problem, the global time reference may be occasionally adjusted by subtracting a constant from all global spike times at the beginning of a layer. The only requirement is that this constant is no larger than the least spike time at the layer's inputs.

8.4   SYNAPTIC WEIGHT TRAINING

Training is performed for each model EC (as implemented in a simulation EC unit). Each EC contains a weight array addressed by input/output pairs, where each element of the array is assigned a weight via training. Training is accomplished in the EC units, and consists of the following steps for each TNN layer to be trained.

1. A volley is retrieved from an addressable buffer that is maintained as a file in secondary storage.

2. The training method described in Section 6.6.2 is implemented. It accumulates a weight increment for each volley

3. After weights for the volleys have been accumulated the EC performs a division to find the averages, which are the final weights. The weights are then stored in a weight array held in secondary storage.

The way training is implemented across multiple layers is to first train TNN Layer 1. After training Layer 1, the training set is applied to Layer 1 a second time and the output volley set is captured in an addressable buffer. This addressable buffer therefore holds the input training set for the next layer. This process is repeated from one layer to the next. At each step, all the ECs in a layer are trained, then the evaluation set is applied to produce the training set for the ECs in the next layer.

8.5   EVALUATION

8.5.1   EC BLOCK

After an EC has been trained, volley sets may be evaluated. This is accomplished via the following steps.

1. Before any input volleys are evaluated, the network weights are read from secondary storage and are then used to generate the E_i functions as described in Section 6.6. The E_i functions are all pre-evaluated and stored in lookup table form. Alternatively, the E_i functions may be evaluated as needed in step (3).

2. An input volley is retrieved from an addressable buffer.

3. P functions are formed and evaluated with respect to the threshold to determine the spike output time, as described in Section 6.6.3.

4. The output volley is formed by combining the output times (values) of the all the neurons making up the EC, and the volley is stored in an addressable buffer.

8.5.2   IC BLOCK

In the simulator, an IC block may implement lateral inhibition, feedforward inhibition, or both.

1. An input volley from an addressable buffer is retrieved.

2. The inhibition function is performed on each of the input volleys, using the k_L and t_L parameters for lateral inhibition and k_F and t_F for feedforward inhibition.

3. The final output volleys are written into the output addressable buffer.

8.5.3   VA BLOCK

The volley analyzer (VA) is a tool implemented in the simulator, but, strictly speaking, it is not part of the actual TNN computation flow. The VA performs volley decoding and analysis as discussed in some depth in Section 7.6. The VA serves a number of functions. One is to produce graphs and tables that provide the designer with feedback regarding the quality of the cognitive processing at any point in the TNN. Another important VA function is to generate the Pareto-optimal data points that lead directly to inhibition parameters.

For any bundle in the network, an important question is: "how much useful clustering information is contained in the volleys being transmitted on this bundle?" This information content can be lower-bounded by employing a table-based decode mechanism contained in a VA.

8.6   DESIGN METHODOLOGY

A TNN architecture is developed one layer at a time starting with Layer 1 which processes input volleys from the RFs. After Layer 1 is designed and trained, then development activity moves to Layer 2. This process continues until all the layers are designed, thereby completing the system.

Each layer is trained and evaluated by executing sequences of MG, VA, IC, and EC models in mostly forward-flow simulation pipelines (Figure 8.3). In Figure 8.3, the flow of spike volleys is shown with heavy dark lines. The flow of control, i.e., the sequence in which the simulation models are executed are shown in normal solid lines. The flow of parameters and weight is shown with dashed lines.

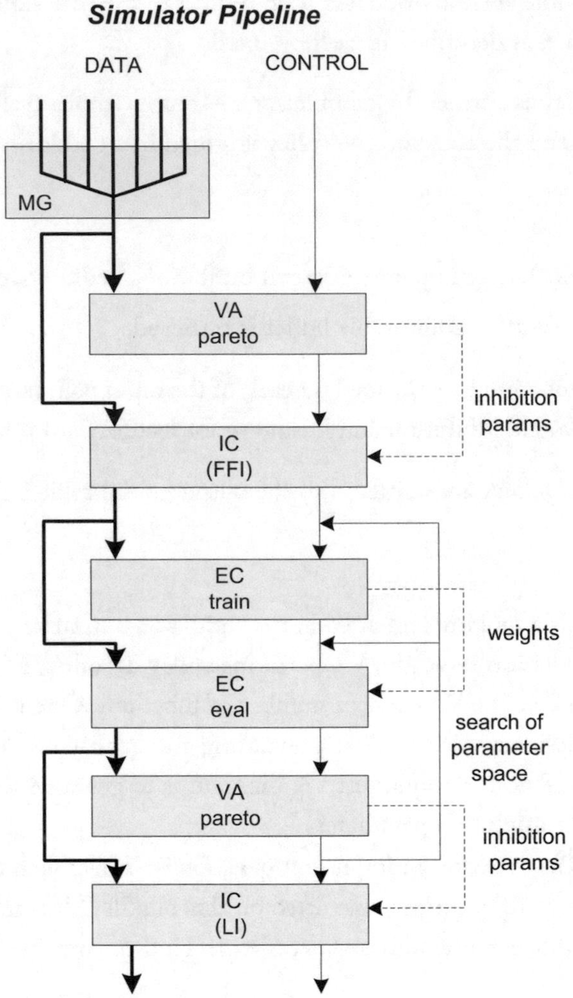

Figure 8.3: Training and testing pipelines for an individual layer. Dark lines indicate flow of data (volley sets); thin lines indicate flow of control; and dashed lines indicate flow of parameters, including synaptic weights.

In general, a layer begins with the merging of volleys from the previous layer, or from the initial receptive fields in the case of Layer 1. This merging into a single volley is done by an MG unit. This single input volley is then analyzed and Pareto-optimal accuracy/coverage points and associated inhibition parameters are generated.

Then, based on the designer's inclination, a point from the Pareto-optimal inhibition set is selected and parameters for the first inhibitory column are determined. The merged input volley then passes through the first inhibitory layer and the volley set (dark lines on left) is saved in an addressable buffer.

Next, the CC is repeatedly trained and evaluated as EC and IC parameters are tuned. Simulation is used for exploring the large design space. A more detailed depiction of loop nesting for this design space search is in Figure 8.4.

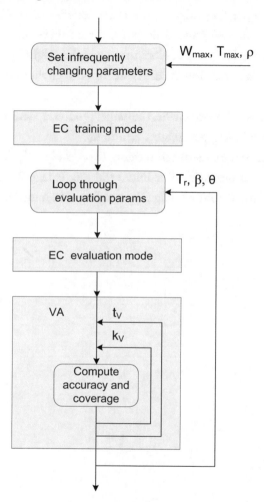

Figure 8.4: Simulation flowchart for design space exploration.

As noted earlier, it is not just the synaptic weights that should be considered plastic. At this stage, several other parameters may be considered plastic. Because the effects of the parameters vary, some are more-or-less fixed during a simulation run, while others are sequenced through a range.

The parameter W_{max} is fixed because changes in W_{max} and the threshold θ directly offset each other; there is no need to vary both.

In the simulation model, T_f is defined in terms of T_r and a proportionality parameter ρ; T_f = $\rho\, T_r$. The value of ρ is generally close to 4, but always > 1. Accuracy is relatively insensitive to ρ. A few initial simulation runs are sufficient for determining a good value to be used throughout.

T_{max} tends to vary depending only on the characteristics of the input volley set. For each layer, it can be established and then fixed. This leaves three inner-loop parameters that effectively define a search space; these are T_r, β, and θ. Note that synaptic weight training is performed outside the inner loops; a single weight training is sufficient for all combinations of the three parameters just given.

After evaluation, a nested loop inside the volley analyzer determines the Pareto-optimal accuracy/coverage points by varying parameters t_V and k_V.

If one implements inhibitory feedback training (Section 7.7.3), then a longer training pipeline is used; see Figure 8.5. In this case, after initial training (Stage 2), the excitatory neurons are trained a second time with the output of Stage 2 providing a training mask.

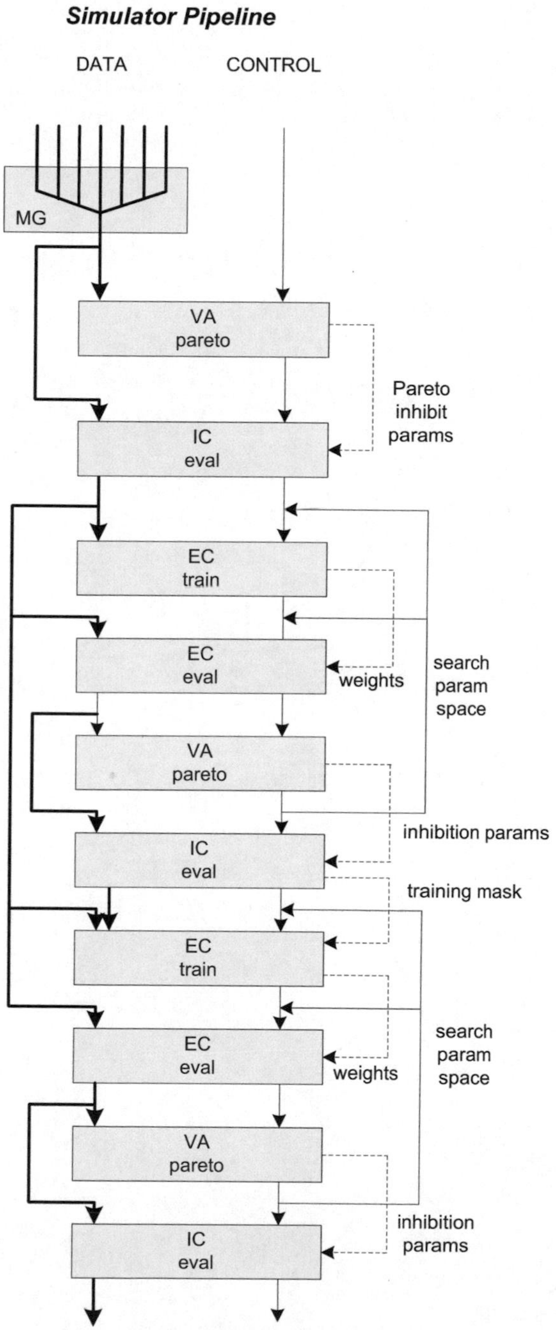

175

Figure 8.5: Simulation pipeline with training mask.

CHAPTER 9

Clustering the MNIST Dataset

As a driver for developing a prototype TNN architecture, the MNIST benchmark [49] provides an excellent workload source. Normally, the MNIST dataset is used for classification. However, in this chapter we focus on clustering of the MNIST dataset via a TNN with unsupervised training.

9.1 MNIST WORKLOAD

The objective of the MNIST benchmark is to take a hand written Arabic numeral "0" through "9" as input and correctly identify, or classify, the numeral that was written. Consequently, the MNIST dataset consists of a large number of Arabic numeral images. Each is 28x28 pixels with grayscale levels from 0–255 (8 bits of grayscale). Figure 9.1 gives some examples of the handwritten numerals. When used as a classifier benchmark, as is most commonly done, the classifier is first trained with a set of numerals from the database (60,000 are prescribed for the benchmark). The classifier is then tested for accuracy using a different set of 10,000 numerals from the database.

In this chapter, we are not developing a classifier *per se*, although the clustering function is the major part of an overall classifier design (briefly discussed in Section 7.6.3). We use the MNIST dataset as the input for clustering operations which involve unsupervised training of excitatory neurons.

Figure 9.1: Some of the handwritten numerals in the MNIST dataset.

9.2 PROTOTYPE CLUSTERING ARCHITECTURE

The general architecture of the prototype TNN is illustrated in Figure 9.2. The inputs to the first layer are receptive fields (RFs). An RF is a subpattern, or "tile" taken from the full pattern, in this case, a 28x28 MNIST handwritten numeral. The full 28x28 pattern is divided into 144 overlapping 5x5 RFs so that adjacent RFs overlap by three pixels. All RFs start on odd numbered rows and columns (denoted by the coordinates of the upper left corner). 144 RFs is a relatively dense tiling, although not the densest possible. (Note that with this particular tiling, the 28th row and column are not covered, so strictly speaking, we are processing pixels from a 27 x 27 image.)

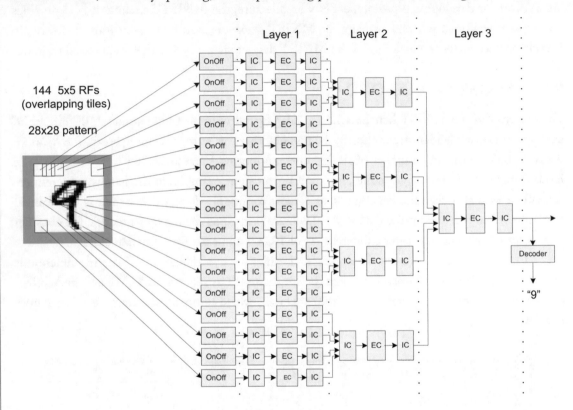

Figure 9.2: Prototype TNN system. Each line represents a bundle carrying a spike volley. This very simple bundle connection pattern is for illustrative purposes. In reality, the interconnect structure is more complex, both between columns and within a column.

As shown in Figure 9.2, the Layer 1 CCs are a special case because Layer 1 must perform the mapping from a purely spatial image to a space-time form. The OnOff block performs OnOff encoding of the RF, doubling bundle size from 25–50 pixels. In Layer 1, information from a single RF is transmitted to an associated EC. In the prototype TNN, each EC consists of 24 neurons and weights are trained without supervision.

To connect Layer 1 outputs to Layer 2 CC inputs, some subset of Layer 1 output bundles are selected for each Layer 2 CC. The interconnection pattern from Layer 1 to Layer 2 may be regular as suggested by Figure 9.2, or it may follow a repeating irregular pattern, or may be pseudo-random.

A key feature of the interconnection pattern is that it incorporates bundle merging followed by EC processing that yields a single output bundle. It is through this merging process that the total numbers of lines are reduced when going from input to output layers.

In Figure 9.2, the CCs in Layer 2 are canonical CCs—the general case. The processing flow is illustrated in Figure 9.3 (copy of Figure 7.2). First, multiple input volleys are merged into a single volley and feedforward inhibition is applied to the merged volley. Then the volley is processed by an EC. Finally, lateral inhibition is applied to the ECs output volleys.

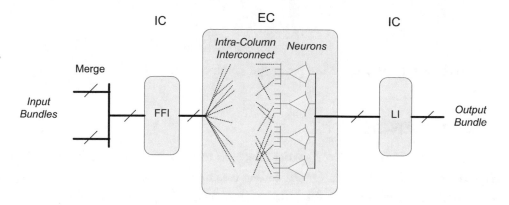

Figure 9.3: Processing flow for generic layer.

The processing and reduction process continues through some number of layers, with Layer 2 serving as a general case template. The last layer shown in Figure 9.2 is Layer 3, only to simplify the figure. In general, there may be more than three layers. In the currently envisaged prototype TNN, the final layer produces a single output bundle consisting of 12–24 lines. The volleys transmitted on this output bundle are decoded into cluster identifiers. These identifiers then access a decode table to match the cluster identifier with a given image label (0–9). The decoding process essentially determines a cluster accuracy (Section 7.6.2) with respect to the MNIST labels.

The Layer 1 study begins with small 5x5 RFs, or "tiles", extracted from the full 28x28 MNIST images. The choice to use 5x5 RFs is more-or-less arbitrary, using deep convolutional neural networks as a guide [20, 88].

Referring to Figure 9.4, the Layer 1 design consists of OnOff coding, a layer of FFI, a pseudo-random interconnection network, an excitatory column, a layer of LI. These major elements will now be discussed in more detail, but will be built up in a different order from the normal input-to-output flow. This is done to illustrate the main features of each more clearly.

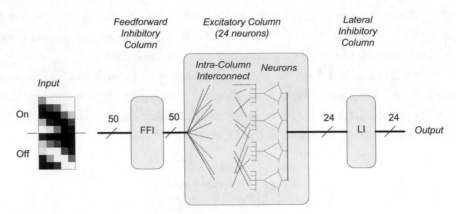

Figure 9.4: Processing flow for Layer 1 of prototype TNN.

9.3 ONOFF ENCODING

The first step in classifying the 28x28 spatial images is to convert them from their original 8-bit grayscale format to spike volleys via OnOff coding (see Section 7.3). As described in that section, the negative of the original pattern is appended to the original. An example of a full MNIST image after OnOff translation is in Figure 9.5. Consequently, a 5x5 RF consists of 50 pixels after encoding.

Figure 9.5: OnOff coding of the Arabic numeral 5.

9.4 INTRA-CC NETWORK

Internal to each CC is an interconnection network that maps a subset of incoming lines to the inputs of excitatory neurons that comprise the EC. A given neuron receives some subset of input lines via this intra-CC network.

With unsupervised training, there must be some variation among the neurons in an EC. If any two neurons are exactly the same: same inputs and same parameters, then they will be trained exactly the same way, they will evaluate exactly the same way, and one of them is essentially redundant.

In the prototype design, it is the Intra-CC interconnection network that provides neuron variation. All other parameters are the same for all neurons in an EC. Consequently, in the current prototype, a pseudo-randomly selected subset of input lines are mapped to each neuron in an EC. After some preliminary experimentation, it was decided that half the lines in the input bundle should map to each neuron. There are 50 lines after OnOff decoding, so each neuron receives a pseudo-randomly selected set of 25 lines. In the baseline design, there are 24 neurons in an EC. Hence, this is denoted as a "50 x 25 x 24" network. The network has 50 inputs, 24 outputs, and each output is connected to 25 of the inputs.

To select a particular pseudo-random interconnection pattern, the following process is followed. First, an arbitrary pseudo-random network is selected. Then, for a particular RF, all the EC parameters are optimized (more details below) for that interconnection pattern. This is only a preliminary optimization. Then 999 other pseudo-random interconnect patterns are generated and each is tested for clustering accuracy with the preliminary EC parameters. Finally, the best performing pseudo-random interconnection pattern is selected. This best-performing interconnection pattern is then used for all the RFs (and ECs) in Layer 1 during subsequent design steps.

9.5   EXCITATORY COLUMN (EC)

As part of a build-up toward a complete CC, we first implement a Layer 1 EC with no inhibition either before or after the EC (Figure 9.6). OnOff encoded 5x5 RF patterns are fed directly into the EC. Then the outputs of the EC are fed into the volley analyzer (decoded) to generate graphs and statistics, including clustering accuracies.

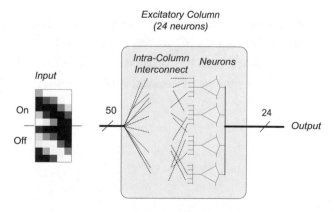

Figure 9.6: Building a CC, Step 1: Apply inputs directly to EC.

As stated earlier, all ECs in a layer use the same non-weight parameters; only the trained weights distinguish the ECs that span all of the RFs. Consequently, we first select one of the RFs for initial EC parameter optimization, and then use the same non-weight parameters in all ECs throughout the layer.

The RF selected for parameter optimization is the one that appears to contain the highest information content, determined here as the RF with the highest clustering accuracy. An initial set of simulation runs indicated that the RF giving the best clustering accuracy is located at <13,15> (i.e., pixels <13:17,15:19>). Figure 9.7a shows the location of the <13,15> RF, and Figure 9.7b shows the contents of the RF for the first 12 MNIST numerals in the benchmark file. These are also the first 12 numerals in the top row of Figure 9.1.

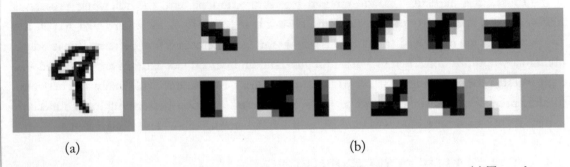

(a)   (b)

Figure 9.7: RF <13,15> is the 5x5 which provides the highest full coverage accuracy. (a) The red square near the center (at <13,15>) indicates a 5x5 RF with high information content. (b) Twelve 5x5 subimages taken from RF <13,15>.

EC Parameters

First, parameters for the excitatory neurons in the EC are optimized. Neuron parameters are listed in the left column of Table 9.1. The first six parameters in the table are for excitatory neurons. The remaining two are inhibition parameters, to be discussed later. The various Table 9.1 columns contain performance-related data for a number of network configurations, to be discussed later.

Table 9.1: Summary of network parameters and performance or 5x5 RF at <13,15>						
		50 x 25 x 24 Network			50 x 25 x 12 Network	
		EC only	EC+LI		EC+LI	
Excitation	T_r (msec)	55	55	55	61	61	61
	ρ	4	4	4	4	4	4
	W_{max}	8	8	8	8	8	8
	T_{max} (msec)	25.6	25.6	25.6	25.6	25.6	25.6
	θ threshold	45	45	45	44	44	44
	β attenuation	.07	.07	.07	.03	.03	.03
Inhibition	t_L (msec)	-	1.5	0.	2.0	1.0	0.
	k_L	-	2	1	4	2	1
Performance	cluster accuracy	.71	.50	.41	.49	.41	.31
	coverage	1.0	1.0	1.0	1.0	1.0	1.0
	cluster count	16723	260	24	693	78	12
	avg. spike count	11.83	1.98	1.0	3.3	1.6	1.0

To optimize EC parameters, the parameter space is swept as per the flowchart in Figure 8.4. The same W_{max} is used throughout the study. It is set at 8 to reflect the assumption that eight simple synapses (maximum weight = 1) are combined into a modeled compound synapse. T_{max} can be varied as part of a parameter space sweep, but it is closely matched to the earliest and latest spike times over a set of volleys, so it is generally fixed throughout the development of a layer. The value of ρ is fixed at 4, so specifying T_r also defines T_f. Consequently, there are only three parameters that are actively adjusted during parameter space exploration. These are the rise time (T_r), the threshold (θ), and attenuation (β).

Overall, the parameter space sweep is managed semi-manually, i.e., simulation currently has active human involvement in the loop. Obviously, the design space exploration can be more automated, however the human-in-the-loop method has been good for building very useful intuition regarding the way TNNs work.

EC Behavior

For the optimized EC at RF<13,15>, Figure 9.8 is a histogram of weights after training. Data is taken from all the synaptic weights across all 24 excitatory neurons. There is a total of 24 * 25 = 600 compound synapses. The vertical bar indicates the number of compound synapses for a given weight. More or less, the weights follow a unimodal distribution centered approximately at 4.

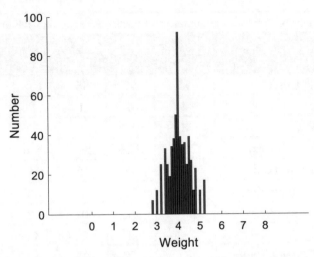

Figure 9.8: Weight distribution for EC at RF <13,15> after training.

After training, the training set is re-applied, and Figure 9.9 contains input and output spike time distributions for the EC at RF <13,15>. Figure 9.9a is the time distribution for input spikes to the EC after OnOff coding is performed. The leftmost and rightmost extremes, which dominate, correspond to pixels that are either white or black in the original image, because OnOff coding translates a black in the original MNIST image to a white in the negative image. This plot clearly shows that we are essentially dealing with black and white images (despite the way MNIST is described, there is very little "gray" in the grayscale). This means that the images are more-or-less temporally "flat" even after translation to spatio-temporal input spike volleys. The very small amount of temporal information shouldn't be ignored, but for temporal processing to be effective in downstream layers, it is important that the ECs in the first layer produce output volleys exhibiting wider temporal variation.

Figure 9.9b shows the spike time distribution for spikes at the EC outputs. Ignoring for the moment the few spike times that occur very frequently, the spikes are reasonably well distributed. There are a few fast spike times (at around 8 msec) that occur frequently, then there is a nice bell shaped curve with a long tail, having a still-significant number of spikes in the 15 msec range. The very late spikes will later be discarded through inhibition. Given that we started with a temporally flat image, the temporal variation at the output of the EC is quite significant, and is what we are striving for.

Given that the first spike in a volley corresponds to the strongest feature, the distribution of first spike times is probably more important than the distribution of all output spike times. Figure 9.9c is the distribution of first output spike times. The distribution is virtually a squashed down version of the distribution for all spikes (note the different y-axis scales in Figure 9.9b and Figure 9.9c.)

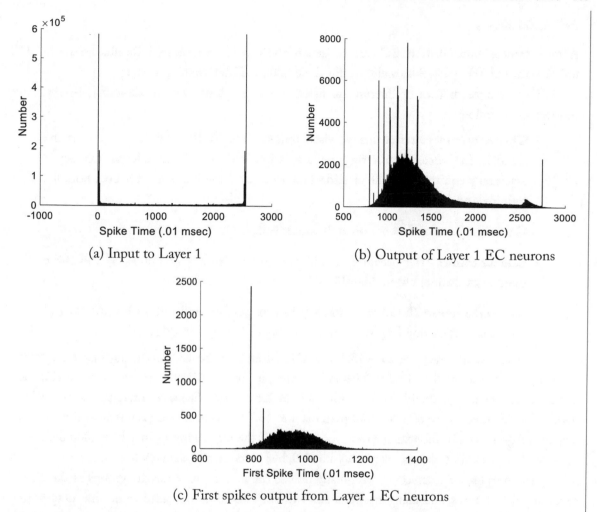

(a) Input to Layer 1

(b) Output of Layer 1 EC neurons

(c) First spikes output from Layer 1 EC neurons

Figure 9.9: Spike time distributions for EC input and output volleys; RF <13,15>.

For both distributions, the very fastest spike time shows about 2,400 occurrences. However, a significant difference in the two graphs is that there are a few fast spikes that occur very frequently in the *all* distribution (Figure 9.9b), but are not present in the first distribution (Figure 9.9c). These missing spikes are 2nd, 3rd, 4th, ... spikes. For the <13,15> RF there is a single input pattern that occurs much more frequently than any other. Given the central location of RF<13,15>, the numeral 0 frequently exhibits an all-white pattern (or nearly so). Hence, a large fraction of numeral 0s are very similar, thereby leading to a large cluster, thereby leading to a particular spiking pattern that occurs very frequently.

EC Performance

After parameter optimization, EC performance for RF<13,15> is measured. Results are in the EC only column of Table 9.1. All results are for the 60,000 MNIST training images.

Performance metrics of interest are listed in the last four rows of Table 9.1. Briefly, the metrics are as follows.

- **Cluster accuracy:** After unsupervised training, the MNIST labels are used as the standard for "correct" clustering; see discussion in Section 7.6.2. Cluster accuracy is essentially equivalent to the classification accuracy if the training set is used both for training and classification.

- **Coverage:** The fraction of non-null output volleys.

- **Cluster count:** After decoding to a binary cluster identifier (Section 7.6), this is the number of distinct cluster identifiers.

- **Avg. spike count:** The average number of spikes per non-null volley. One can arrive at a *spike density* by dividing by the total number of lines (24 in this case).

The cluster accuracy for RF <13,15> is .71. Given that the EC producing this accuracy is peering through a small 5x5 pixel "peephole" (examples of what it sees are in Figure 9.7b), the .71 accuracy seems remarkably high. However, if we look at the cluster count, enthusiasm quickly fades. The EC divides the 60,000 input patterns into 16,723 clusters. There is relatively little actual clustering going on. In the extreme case, for example, if the 60,000 inputs map to 60,000 different cluster identifiers, then the accuracy would be 1.0, but no clustering actually occurs.

The purpose of clustering is to greatly reduce the number of similarity groups as one goes from input to output, and using an EC alone without inhibition has failed to do this, despite its apparently good accuracy. This problem can be alleviated by adding lateral inhibition to the EC output in order to greatly improve the amount of clustering.

9.6 LATERAL INHIBITION

The cluster count is the number of different binary patterns (cluster identifiers) after decoding. If we begin with 24 excitatory neurons, and place no limits on the particular binary patterns, as was done in the "EC only" system, we wind up with a large number of densely coded cluster identifiers. On average, half the 24 output lines contain a spike (*avg. spike count* = 11.83 in Table 9.1).

The number of clusters can be greatly reduced based on the principle that the first (earlier) spikes are the most important, i.e., they identify the strongest features. Consequently, we use lateral inhibition (LI) to eliminate all but the very early spikes. This results in sparse cluster identifiers and a much smaller cluster count. After adding LI, the CC under construction is shown in Figure 9.10.

It was decided to limit the output spikes to the first 2 in each volley via LI. Hence, the parameter k_L is set to 2. Then, t_L is varied over a range to determine the t_L that yields the best cluster accuracy. For the design we are working on, both k_L and t_L are given in the leftmost "EC + LI" column of Table 9.1, along with peformance results. k_L and t_L are 2 and 1.5 msec. What this means is that inhibition first restricts consideration to a 1.5 msec time window following the first spike, and then takes the first 2 spikes within that window. This may occassionally result in only a single spike, when the second spike occurs more than 1.5 msec later than the first.

Refering to the performance data, after LI is added to the EC, we see that the accuracy is reduced to .50, which still seems quite good for a single 5x5 RF, and, very importantly, the number of clusters is reduced dramatically to 260. The spike density, as one might expect, is reduced to about 2 per 24 line volley.

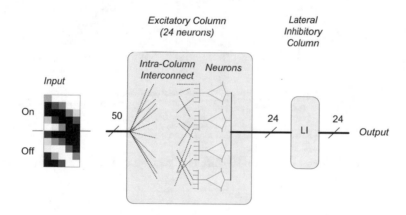

Figure 9.10: Building a CC, Step 2: Add lateral inhibition to output of EC.

The performance of EC + LI was tested with other configurations, some of which are shown in the right-hand columns of Table 9.1.

A more complete set of data are plotted in Figure 9.11. Notation: *M/N* denotes a system with *M* excitatory neurons (outputs), and k_L = *N*. This data shows the relationships between neuron counts and k_L, and cluster count and accuracy. First, as noted in the figure, better data points are down (fewer clusters) and to the right (better accuracy). In general, the data points with 24 neurons are better. They are not only better in terms of accuracy and cluster count, they are also better in terms of spike sparsity.

Hence, both 24/1 and 24/2 might be good choices. After some thought, the 24/2 system was selected for further study. One obvious advantage of 24/2 is its better accuracy versus 24/1 and, ultimately, we want high accuracy. A second advantage is that using two spikes allows much more temporal information to be encoded in a volley (it is important to recall that this temporal information is discarded when computing cluster counts). Finally, purely from a research perspective, study-ing two spikes per volley seems more general (and more interesting) than a single spike per volley.

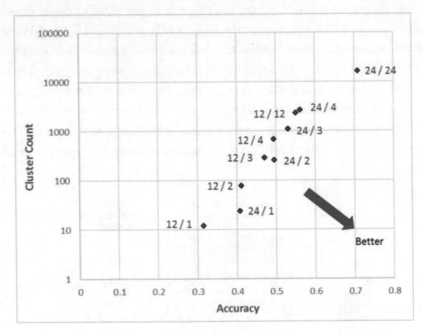

Figure 9.11: Accuracy and cluster count for 10 different "neuron count / k_L" configurations

9.7   144 RFS

Using the neuron parameters optimized for RF<13,15> : (Table 9.1, EC + LI, 50 x 25 x 24 network, k_L = 2), a spatial sweep of all the overlapping 144 RFs was performed and the clustering accuracies are summarized in Figure 9.12. For each RF the clustering accuracy is given, and a colored "heat map" illustrates accuracy as a function of RF position.

As shown by the accuracy heat map, the better accuracies are in the more information-rich center of the full image, with .50 being the best as given earlier. Note the slight slant to the hot spots that reflects the slight slant that tends to occur when a right-handed person writes numerals. This suggests that the volley sets with higher information content roughly follow the slant. Around the center of the image, the accuracies are quite good (given that they represent accuracy extracted from a single 5 x 5).

	1	3	5	7	9	11	13	15	17	19	21	23
1	.11	.11	.12	.15	.17	.20	.22	.23	.20	.16	.13	.12
3	.11	.13	.16	.20	.23	.29	.32	.34	.31	.25	.18	.15
5	.12	.15	.19	.26	.34	.38	.41	.42	.39	.32	.24	.18
7	.13	.16	.22	.31	.37	.42	.42	.44	.41	.36	.27	.21
9	.14	.17	.23	.32	.39	.43	.39	.41	.39	.35	.28	.22
11	.13	.17	.24	.35	.42	.43	.42	.45	.41	.36	.30	.22
13	.13	.18	.26	.38	.44	.44	.48	.50	.42	.37	.28	.20
15	.13	.20	.28	.36	.43	.41	.44	.43	.39	.36	.28	.21
17	.14	.20	.25	.35	.41	.38	.39	.38	.38	.33	.26	.21
19	.14	.20	.23	.33	.39	.39	.37	.36	.34	.29	.23	.18
21	.13	.17	.21	.29	.37	.38	.36	.31	.29	.24	.18	.15
23	.12	.14	.17	.22	.27	.31	.29	.24	.21	.18	.15	.13

Figure 9.12: Clustering accuracies for 144 overlapping 5x5 RFs, arranged over a 28x28 MNIST image. A heat map highlights relative accuracy.

As one considers RFs farther from the <13,15> center, the accuracies decrease. The RFs on the periphery of the 28x28 images appear to be relatively information poor (they are typically all white or nearly all-white 5x5s). For example, in the upper left corner, the accuracies are .11, which is slightly higher than the .1 accuracy that random chance might suggest. The reason for this slight difference is that all 10 numerals are not equally represented in the 60 K images in the dataset. The numeral that appears most often (the numeral 1) results in the .11 accuracy RFs on the periphery. Essentially, for these RFs, the decoding method naturally degenerates toward always predicting the most commonly occurring numeral over the entire dataset.

9.8   FEEDFORWARD INHIBITION

Given the observation just made in the previous section, one might conclude that there is very little useful information along the periphery of an image and let it go at that. This is not the case. Applying feedforward inhibition (FFI) to the input volleys reveals a significant amount of useful information; see Figure 9.13.

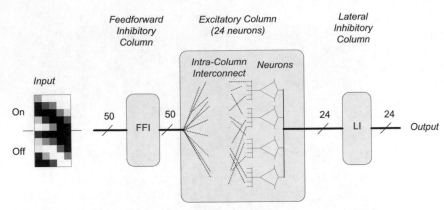

Figure 9.13: Building a CC, Step 3: Add feedforward inhibition to input of EC.

Recall the brief discussion in Section 7.5 concerning the filtering effect of inhibition. The simple intuition is that if all feature values are equally high (as indicated by early simultaneous spikes), there is little or no useful information for separating the features. This large volume of apparently useless information can be converted into a smaller amount of useful information via FFI.

With FFI, any input volley having more than k_F spikes within time t_F of the first spike is nullified by converting it to a volley having no spikes at all. After all such volleys containing minimal useful information are removed, the remaining non-nullified volleys are relatively information rich. However, in exchange for the information richness, the coverage may be significantly less than 1. In general, for a given RF, the higher the coverage, the lower the accuracy and vice versa (Figure 7.9).

Training Inhibition

Volley analysis (VA) is an integral part of the inhibition training process. As discussed in Section 7.7.1, by allowing VA parameters t_F and k_F to range over a two dimensional parameter space, one can compute an accuracy and a coverage for each parameter pair. Then, from the full set of accuracy/coverage data points, one can construct a Pareto-optimal set. As part of the IC design process, these Pareto-optimal data points are determined for all 144 RFs. This is relatively easily done because the computing the clustering accuracy for a t_F, k_F pair is very fast for a given set of volleys.

In the prototype TNN, the overall FFI tuning method strives for a similar accuracy across all the RFs (144 in the case at hand), and, provided at least this accuracy achieved, the tuning method then strives for the highest coverage. For an *iso-accuracy* of .4, for example (see Section 7.7.1), one selects the lowest accuracy on each of the Pareto-optimal curves that is at least .4. Or, if there is no such point, the highest accuracy is selected, regardless of the coverage. At these selected data points, we get the highest coverage we can get for the desired accuracy (by definition of "Pareto-optimal").

Example

To illustrate the overall process of training FFI, the iso-accuracy point of .68 is chosen for the input RFs. Then, consider RF <7,7> as an arbitrarily chosen example.

Volley analysis is applied to all the volleys in the training set for RF<7,7>. VA performs an accuracy/coverage analysis for a range of t_F, k_F pairs. In this example, t_F varied from .1 msec to 2.0 msec in increments of .1 msec. k_F varied from 1 to 24 (the maximum number of spikes in the volleys being analyzed). For each <t_F, k_F > pair, an accuracy and coverage is first computed. Then, the set of pairs is reduced to the set of Pareto-optimal points (Figure 9.14). If we assume iso-accuracy of .68, the selected Pareto-optimal point is shown in Figure 9.14. That is, an accuracy of .68 with coverage .11 can be achieved with t_F = 1.3 msec and k_F = 10, i.e., if there are more than 10 spikes within 1.3 msec of the first spike, then the volley is nullified. □

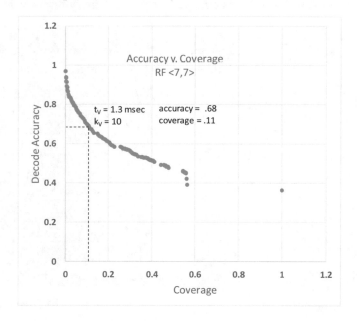

Figure 9.14: Pareto optimal data points for RF <7,7>.

Returning to the prototype TNN, this process of finding optimal FFI parameters was done for all 144 RFs, with Pareto-optimal parameters t_F and k_F being determined for each. An *iso_accuracy* of .7 was eventually settled on. Observe that FFI is applied to the inputs before they are processed by the EC, so their accuracies (and cluster counts) are relatively high. Then, after FFI is applied to all 144 RFs, the results at the CC outputs, both accuracies and coverages, are given in Figure 9.15. The coverages are multiplied by the number of images (60 K), giving the number of images covered at each RF.

These results should be compared with the results in Figure 9.12, which are produced without FFI at the inputs. Note that in Figure 9.12, the coverage is 1.0 for all RFs. Comparing the accuracies in the two figures, we see that FFI has the desired effect. In Figure 9.15 near the center of the image, where the 5x5 RFs are relatively information rich, we get the same accuracies as for EC + LI alone (Figure 9.12) with 1.0 coverage in both cases. On the periphery, however, the results are now dramatically different with FFI than without. With FFI, we now achieve relatively high accuracies for many of the RFs near the periphery.

As expected, these high accuracies come at the expense of lowered coverages. This is a good tradeoff, however. Without inhibition, we received little or no useful information from the RFs on the periphery, even though the ECs at these RFs emit huge numbers of spikes (prior to LI). With FFI, almost all of the spikes are eliminated as volleys are nullified, and the remaining non-null volleys contain very useful information as reflected by their relatively high clustering accuracies.

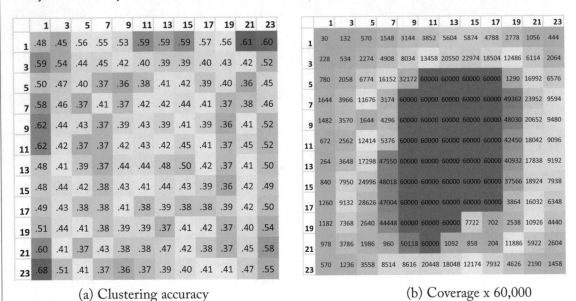

(a) Clustering accuracy (b) Coverage x 60,000

Figure 9.15: Accuracy and coverage after FFI is implemented. Coverage is given as the number of non-null images.

Consider the extreme case of RF <3,1>, i.e., in the upper left corner. The accuracy is .59. In this case, exactly 228 of the 60 K images are covered. That is, this RF rarely yields any EC output spikes, but when it does, the spikes provide high quality information. The reason is that this part of the image is *almost* always white space. All volleys containing only white space are nullified. We are then left with the very few images where this RF is not completely white space.

The first twelve of the covered 28x28 images are shown in Figure 9.16. The top two rows contain the full images, the next two rows contain only the 5x5 RF <3,1> subimages. The majority of the images are in the cluster for correct label "3." The upper-left tip of these few 3's impinge on the lower-right corner of the <3,1> RF. Four of the incorrect cases are 2's and 4's that also impinge on the lower-right corner of the RF. The remaining incorrect case is a "1" where there apparently was an extraneous spot on the original scanned image; perhaps a speck of dust. This spot is detected via FFI. Presumably when the output of the EC operating on this column is combined in later clustering layers with results from other RFs, such extraneous cases will be easily over-ridden.

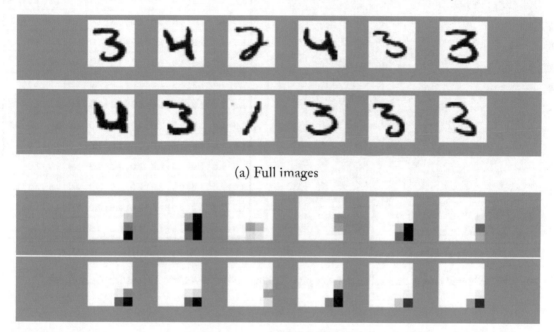

(a) Full images

(b) 5x5 subimage at RF <3,1>

Figure 9.16: First 12 non-null images for RF <3,1>.

The implications of this inhibition-invoked behavior are promising. In particular, it illustrates a potential benefit of FFI. The input patterns that will make-or-break a clustering method are not the easy ones, but the hard ones. That is, the ones with odd features and quirks that often are most apparent around the periphery. FFI focuses attention, so to speak, on these.

Finally, it is informative to consider the cluster counts at each of the RFs; see Figure 9.17a. The cluster counts for RFs near the center of the image are relatively high, approximately 250–270 decoded clusters. This is the more information rich region of the image, so one would expect that more clusters would be required. Meanwhile, closer to the periphery, the number of clusters is much lower, less than 100 in many cases. The number of clusters naturally adjusts to the data that is being analyzed.

Regarding spike density (or sparsity), Figure 9.17b shows that in most cases there are 2 spikes per 24 output lines from the EC. This is expected because $k_F = 2$ limits the number of spikes to at most two. Around the edges, this number sometimes dips closer to one, indicating that better accuracy/coverage points are achieved via cluster identifiers having a single spike.

	1	3	5	7	9	11	13	15	17	19	21	23
1	2	7	16	28	40	54	68	68	62	56	43	24
3	7	16	41	75	121	146	164	165	149	113	90	64
5	12	38	87	140	183	207	250	258	238	21	160	97
7	26	55	137	23	233	257	275	272	266	238	190	108
9	25	56	20	23	258	270	270	267	257	242	191	100
11	22	65	166	24	256	264	269	267	250	236	189	95
13	17	61	176	230	258	261	272	260	253	227	184	99
15	15	65	171	251	261	279	276	269	258	227	192	102
17	21	74	189	238	275	270	276	266	236	22	176	91
19	23	70	23	229	245	261	262	23	19	22	143	72
21	16	44	16	20	227	241	16	17	16	150	88	44
23	12	25	56	99	19	155	158	144	128	78	45	25

(a) Cluster count

	1	3	5	7	9	11	13	15	17	19	21	23
1	1.1	1.6	1.2	1.4	2.0	2.0	2.0	2.0	2.0	1.9	1.6	1.8
3	2.0	2.0	2.0	2.0	2.0	1.8	1.8	1.9	1.8	2.0	2.0	1.5
5	1.3	2.0	2.0	2.0	1.9	1.9	1.9	1.8	1.8	2.0	2.0	1.9
7	1.8	2.0	2.0	1.9	2.0	1.9	1.9	2.0	2.0	2.0	2.0	1.9
9	2.0	1.4	1.9	1.9	1.8	2.0	1.8	2.0	1.8	1.8	2.0	1.9
11	1.2	2.0	1.8	1.9	1.8	1.8	2.0	1.8	1.8	2.0	2.0	1.6
13	1.9	2.0	1.9	2.0	2.0	2.0	2.0	2.0	1.8	2.0	1.6	2.0
15	1.9	2.0	1.9	1.8	2.0	2.0	1.8	2.0	1.8	2.0	1.9	1.9
17	2.0	1.6	1.9	2.0	1.7	2.0	2.0	1.9	1.8	1.9	2.0	1.9
19	2.0	2.0	1.9	2.0	2.0	2.0	1.9	1.9	2.0	1.9	2.0	1.9
21	1.1	1.3	1.9	2.0	1.9	1.8	2.0	2.0	2.0	1.9	1.9	1.6
23	2.0	1.9	2.0	1.8	1.5	1.9	1.9	1.9	2.0	1.5	1.6	2.0

(b) Average spikes per volley

Figure 9.17: Cluster count and spike density (avg. spikes per volley) after FFI is implemented.

9.9 LAYER 1 RESULT SUMMARY

Layer 1 is crucial because it converts spatial input patterns into the space-time raw material used by all the subsequent clustering layers. In addition to space-time coding, Layer 1 also performs a modest amount of clustering.

A Layer 1 EC followed by LI reduces the spatially coded 5x5 RF input to a sparse temporal encoding transmitted on a 24-line bundle. There is enough information in the output volley of the EC to allow clustering with a .45+ accuracy with 100% coverage for the better-case RFs near the center of the input 28x28 grayscale pattern. For RFs away from the center, the importance of inhibition becomes apparent. FFI reduces the total number of spikes dramatically, with the accuracy provided by the remaining spikes greatly improved (albeit with reduced coverage).

The complexity for training is extremely low; basically, it involves computing average spike times over the spikes in the volley. A single pass through the training dataset is sufficient.

As of this writing, Layer 1 is as far as the detailed design effort has progressed. The next section is a case study of a related architecture to provide some perspective. This is followed by a subsection discussing issues related to future Layer 2 design.

9.10 RELATED WORK

To provide some perspective, this section describes prior work by Masquelier and Thorpe [66], which has recently been enhanced by Keradpisheh, Ganjtabesh, and Masquelier [47]. The overall architecture is inspired by earlier work of Serre, Wolf, and Poggio [87], although this earlier architecture does not use spiking neurons.

This particular architecture [47] is chosen for discussion because it is organizationally similar to the architecture proposed here. It is a TNN employing spiking neurons which operate on a single wave of spikes that propagates forward through the network. However, it also differs from the architecture proposed in this book in significant ways that provide some interesting contrasts.

The overall architecture is shown in Figure 9.18. This architecture is targeted specifically toward vision. Although we have worked with grayscale images in this chapter and elsewhere, we are not performing vision *per se*. Raw grayscale images are treated as surrogates for any of the general class of numerical patterns, and we do not take advantage of features specific to visual objects.

When designing for vision specifically, one can apply effective edge-detecting filters to the input images. These extracts the more useful information (edges) and puts it into a temporal encoding. This is in contrast to the very simple On-Off encoding used in this work. With On-Off encoding, there is much less information pre-processing than with more sophisticated edge detecting filters.

The architecture in [47] and [66] is also more advanced with respect to orientation and scale invariance. The authors/designers use separate maps for each of four different image orientations and five different scales. The different orientations (but not the different scales) are shown in S1 in the figure.

From S1 to C1, winner-take-all lateral inhibition is applied. S2 contains a layer of excitatory integrate-and-fire neurons. These neurons are trained via STDP and do not leak. The given justification for non-leaking neurons is that they are reset between input applications, anyway. In effect, it is assumed that biological leaking is present only to provide a reset for the next input pattern, rather than being part of processing the current input pattern.

The C2 layer is basically a second layer of winner-take-all inhibition. The final outputs of C2 are fed into a conventional machine learning classifier to arrive at the final output classification.

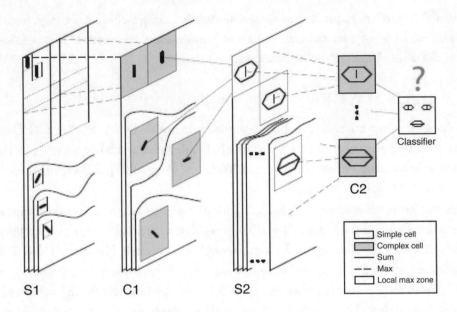

Figure 9.18: TNN architecture used in [47] and [66]. In S1, edge detecting filters are applied at different orientations and scales. This is followed by inhibition in C1, excitation in S2, inhibition in C2, and final classification at the output.

The S1, C1, S2, and C2 layers basically perform encoding, inhibition, excitation, and inhibition, respectively. Although they are termed as different layers in [47] and [66], they are a single layer (Layer 1) in the terms used in this book.

Following are significant differences with respect to the detailed TNN architecture studied throughout this chapter.

1. It allows for images with different orientations and scales to provide invariance. The simple MNIST benchmark has a single scale and orientation.

2. It employs more sophisticated pattern encoding (edge-detection) which is targeted at vision.

3. The first layer of inhibition is lateral, not feedforward as in the design in this chapter. The architecture in [47] and [66] does not appear to use feedforward inhibition.

4. Neurons do not leak. Furthermore, they have only simple synapses, not the compound synapses which are a foundation of the work in this chapter. Another difference is that the excitatory neurons employ weight sharing; that is, the same weights are replicated in all the S2 scale maps.

5. It uses a more sophisticated machine learning classifier at the final output. The neural network is essentially complete with only Layer 1 (in the terms used here). In contrast, the approach used in this book is to take the Layer 1 outputs and feed them to additional neuron layers until a very simple classification method is achieved.

9.11   CONSIDERING LAYER 2

Returning to the design at hand, the outputs of Layer 1 CCs will provide the raw material for Layer 2; see Figure 9.15. Many of the Layer 1 columns achieve an accuracy of at least .4. If one assumes independence amongst all the 144 column "predictions", the overall accuracy becomes extremely high. (The "prediction" is the most frequently-occurring label for each cluster identifier). Unfortunately, the columns are far from independent, but the basic principle holds: multiple predictions taken from different spatial locations will combine to yield a much higher overall accuracy.

Consider two specific MNIST images in Figure 9.19: images 11 (labeled "3") and 12 (labeled "5"). Both the grayscale image and the most common label for each of the overlapping 144 RFs is shown. First note that all the white areas produce no label; these correspond to null volleys after the initial FFI in Layer 1.

For both images, the plurality of the labels is correct. However, considering only the plurality disregards a lot of spatial information. For example, the bottom half of a 3 and the bottom half of a 5 are very similar. This shows up for both images. One of the RFs near the bottom of the 3 image produces the label "5". Meanwhile, a few of the RFs near the bottom of the 5 produce the label "3". In both cases, the alternate label occurs frequently, it's just not the *most* frequent. If one were to characterize a more accurate label for the associated cluster identifiers it would be more like "lower half 3 or 5". Then, when combined with spatial upper half information, where the 3 and 5 are quite different, a definitive identification can be made.

While considering these images, note that the 5 is rather strangely written—with a long scrawly "tail" in the upper-right corner of the image. In the upper-right corner, the RFs are strongly labeled "5". The reason is that a handwritten 5 is the most common numeral that impinges on this peripheral area. The coverage in this area is less than .20, but the accuracies are about .5.

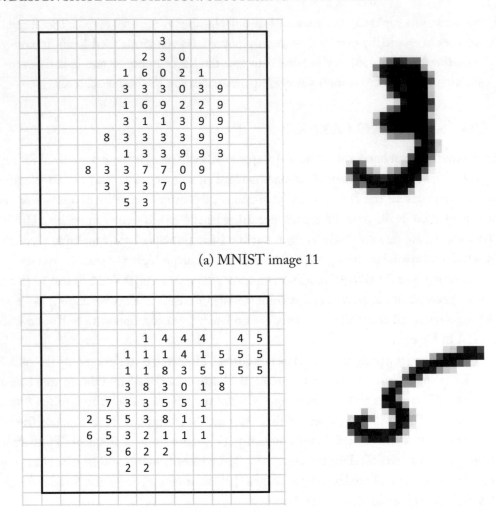

(a) MNIST image 11

(b) MNIST image 12

Figure 9.19: Two MNIST images and the corresponding RF "predictions" (most common label).

In general, the design space for Layer 2 is significantly richer than for Layer 1. For one thing, the input volleys to Layer 2 have much more temporal variation (compare Figure 9.9a with Figure 9.9b and Figure 9.9c). The input volleys to Layer 2 will also be much sparser than the inputs to Layer 1.

As Layer 2 results are developed, they may be fed back into a re-design of Layer 1 using other configurations and parameters at that layer. For example, the 144 5x5 RFs result from an overlap of three pixels between adjacent RFs. One could also use, for example one pixel overlaps, which would result in significantly fewer Layer 1 RFs. Furthermore, RFs with even-numbered (row, column) coordinates can also be considered. The choice of 5x5 RFs was somewhat arbitrary and based only some preliminary study. One could just as easily use 4x4s, 6x6s, or something larger or smaller. However, such design decisions can best be made in the context of an overall design, so the design of Layer 2 will very likely lead to some re-design and additional design space exploration of Layer 1.

Finally, there are a number of interconnection structures to be considered. The three schematics in Figure 9.20 illustrate three potential ways of constructing a Layer 2 CC by combining the input bundles from a selected set of Layer 1 CCs. The network illustrated in Figure 9.20a essentially draws from a field in the original image by combining four non-overlapping 5x5 RFs. The network illustrated in Figure 9.20b combines nine overlapping 5x5 RFs to cover the same area as four non-overlapping RFs. The network illustrated in Figure 9.20c combines a more general pattern set of 5x5 RFs. These may be selected systematically or pseudo-randomly.

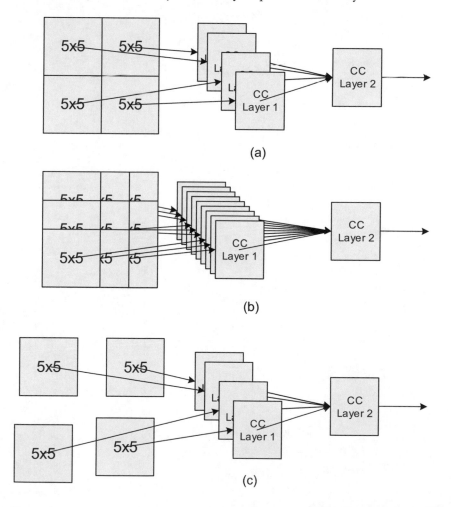

Figure 9.20: Example two layer configurations which map aggregate RFs from the original image.

CHAPTER 10

Summary and Conclusions

In this book, four main assertions are put forward.

1. *In developing the neocortex, evolution used time as a communication and computation resource.*

   As a physical resource, time is freely available. It consumes no power and takes up no space. The usefulness of time for communication and computation was illustrated near the beginning of this book via simple examples employing spikes transmitted over wires with transmission delays.

   To support evolution's use of time as a resource, there is plenty of experimental evidence that volleys of biological spikes have precisely maintained timing relationships—a necessary property. Going further, an argument supporting the point is: if evolution created neurons that can maintain such tight precision, then there must have been a reason. And that reason may well have been communication and computation.

   Perhaps the best experimental evidence for evolution using time as a resource is that volleys of information-carrying first spikes are processed by feedforward sensory pathways—pathways that must be very fast because they are key to an organism's survival in the natural world.

   There appears to be solid evidence that this first point is valid. At least there is enough evidence to proceed further, based on this premise.

2. *This led to a class of computing (cognitive) paradigms that are fundamentally different from the ones that we (humans) tend to devise.*

   When we design a computing paradigm, we have shown a strong tendency to use paradigms where the passage of time in the physical implementation does not affect the results. This is the case in all of today's commercial computers. When a logic network is implemented in silicon, it consists of ranks of clocked flipflops interleaved with combinational logic networks. The clock cycle is determined by worst-case transistor delays and path lengths. Then when fabricated in silicon, computed results will be the same for different chips, despite any variations in transistor and wire delays among the chips. One could shift transistor delays in a given chip, and as long as the worst-case path constraints are satisfied, the results will not change. Results are independent of physical time.

In contrast, if one were to use the passage of physical time as a communication and computation resource using precisely timed spikes, then results become highly dependent on the time it takes to compute (and communicate) them. In fact, when all information is communicated and processed as spikes in time, the time it takes to compute a value is the value. For example, if one were to start with a correctly functioning section of biological neocortex, and then shift communication delays and neuron computation latencies, even by single milliseconds, that section of neocortex would stop working in all likelihood. (At least until a stable operating point could be re-established.) Results are dependent on physical time.

3. *This class of computing paradigms can be clearly characterized.*

The paradigms fit into the general class of space-time computing systems. As defined here, these systems have causality and invariance as their two main properties. The individual primitive functions that comprise a space-time system can be any computable function that exhibits both causality and invariance—not just functions based on biological neurons.

This broad class of space-time computing systems is isomorphic to TNNs, as defined here. This means that the "neurons" in a TNN can compute any computable function. And that means that TNNs include all possible implementable spiking neuron models that are causal and invariant.

4. *Researching this class may lead to entirely new ways of building computers that have advanced cognitive capabilities.*

In a way, this point is a conclusion from the other three. The bulk of this book is research into the class of space-time computing systems, in the guise of biological-ly-based TNNs. The hope is that this approach will lead eventually to entirely new ways of building computers.

Because we are using the biological neocortex as a guide, we would like to construct a bridge between the biology and the basic features of TNNs. In a TNN, processing is done in a feedforward manner as a wide volley of spikes sweeps from the inputs, through layers of excitation and inhibition, to the TNN's outputs. This seems to be a far cry from the way the biological neocortex appears.

However, the gap can be bridged. We start with a spiking neural network model based on widely accepted first-order concepts and features: LIF neurons that communicate via spikes, STDP, multiple inter-neuron paths with variable delays, and bulk inhibition. Then, a series of modeling steps and simplifications are applied. Feedforward networks are shown to be computationally equivalent to feedback networks. Spike trains are divided into bursts by inhibitory oscillations. The first spike in each burst carries all the information, so the other spikes can be eliminated. We are left with a volley of first spikes that sweeps through a feedforward network—a TNN, in other words.

By using a concurrently developed simulator, a prototype TNN is under development with the objective of accurately clustering grayscale input patterns (the MNIST database). However, it is important to note that grayscale images are good surrogates for a broad class of inputs: two dimensional arrays composed of ordered values.

Although far from complete, the design study is targeted at scalable, hierarchical, multi-layer architectures incorporating column-based computation, where excitation and inhibition are integrated within each computational column. Both spatial to space-time encoding and decoding are implemented. As expected, unsupervised training is as fast as evaluation (faster, actually). Thus far, only Layer 1 of a multi-layer TNN has been fully developed. This first layer operating on 5x5 RFs not only yields good accuracies, but the neuron models appear to operate internally as expected.

References

1. Arthur, J. V., P. A. Merolla, F. Akopyan, R. Alvarez, A. Cassidy, S. Chandra, S. K. Esser, N. Imam, W. Risk, D. B. Rubin, and R. Manohar. "Building block of a programmable neuromorphic substrate: A digital neurosynaptic core." *The 2012 International Joint Conference on Neural Networks* (IJCNN) (2012): 1–8. DOI: 10.1109/ijcnn.2012.6252637. 44, 106

2. Bakkum, Douglas J., Zenas C. Chao, and Steve M. Potter. "Long-term activity-dependent plasticity of action potential propagation delay and amplitude in cortical networks." *PLOS One* 3.5 (2008):e2088. 62

3. Bi, Guo-qiang, and Mu-ming Poo. "Synaptic modifications in cultured hippocampal neurons: dependence on spike timing, synaptic strength, and postsynaptic cell type." *Journal of Neuroscience* 18, no. 24 (1998): 10464–10472. 50, 51, 108

4. Beierlein, Michael, Jay R. Gibson, and Barry W. Connors. "Two dynamically distinct inhibitory networks in layer 4 of the neocortex." *Journal of Neurophysiology* 90.5 (2003): 2987–3000. DOI: 10.1152/jn.00283.2003. 53

5. Beyeler, Michael, Nikil D. Dutt, and Jeffrey L. Krichmar. "Categorization and decision-making in a neurobiologically plausible spiking network using a STDP-like learning rule." *Neural Networks* 48 (2013): 109–124. DOI: 10.1016/j.neunet.2013.07.012. 18

6. Bichler, Olivier, Damien Querlioz, Simon J. Thorpe, Jean-Philippe Bourgoin, and Christian Gamrat. "Extraction of temporally correlated features from dynamic vision sensors with spike-timing-dependent plasticity." *Neural Networks* 32 (2012): 339–348. DOI: 10.1016/j.neunet.2012.02.022.

7. Bohte, Sander M., Han La Poutré, and Joost N. Kok. "Unsupervised clustering with spiking neurons by sparse temporal coding and multilayer RBF networks." *IEEE Transactions on Neural Networks* 13, no. 2 (2002): 426–435. DOI: 10.1109/72.991428. 103

8. Bohte, Sander M., Joost N. Kok, and Han La Poutré. "Error-backpropagation in temporally encoded networks of spiking neurons." *Neurocomputing* 48, no. 1 (2002): 17–37. DOI: 10.1016/S0925-2312(01)00658-0. 103

9. Brader, Joseph M., Walter Senn, and Stefano Fusi. "Learning real-world stimuli in a neural network with spike-driven synaptic dynamics." *Neural Computation* 19, no. 11 (2007): 2881–2912. DOI: 10.1162/neco.2007.19.11.2881. 18

10. Bucher, Dirk and Jean-Marc Goaillard. "Beyond faithful conduction: short-term dynamics, neuromodulation, and long-term regulation of spike propagation in the axon." *Progress in Neurobiology* 94.4 (2011): 307–346. DOI: 10.1016/j.pneurobio.2011.06.001. 88

11. Budd, Julian M. L. and Zoltán F. Kisvárday. "Communication and wiring in the cortical connectome." *Frontiers in Neuroanatomy* 6 (2012): 1–23. 62

12. Bullock, Theodore H., Michael V. L. Bennett, Daniel Johnston, Robert Josephson, Eve Marder, and R. Douglas Fields. "The neuron doctrine, redux." *Science* 310, no. 5749 (2005): 791–793. DOI: 10.1126/science.1114394. 4

13. Butts, Daniel A., Chong Weng, Jianzhong Jin, Chun-I. Yeh, Nicholas A. Lesica, Jose-Manuel Alonso, and Garrett B. Stanley. "Temporal precision in the neural code and the timescales of natural vision." *Nature* 449, no. 7158 (2007): 92–95. DOI: 10.1038/nature06105. 72

14. Buxhoeveden, Daniel P. and Manuel F. Casanova. "The minicolumn hypothesis in neuroscience." *Brain* 125, no. 5 (2002): 935–951. DOI: 10.1093/brain/awf110. 60

15. Cardin, Jessica A., Marie Carlen, Konstantinos Meletis, Ulf Knoblich, Feng Zhang, Karl Deisseroth, Li-Huei Tsai, and Christopher I. Moore. "Driving fast-spiking cells induces gamma rhythm and controls sensory responses." *Nature* 459, no. 7247 (2009): 663–667. DOI: 10.1038/nature08002. 77

16. Carlo, C. Nikoosh and Charles F. Stevens. "Structural uniformity of neocortex, revisited." *Proceedings of the National Academy of Sciences* 110, no. 4 (2013): 1488–1493. DOI: 10.1073/pnas.1221398110. 77

17. Casasanto, Daniel and Lera Boroditsky. "Time in the mind: Using space to think about time." *Cognition* 106, no. 2 (2008): 579–593. DOI: 10.1016/j.cognition.2007.03.004. 27

18. Chaitin, Gregory. *Meta Math!: The Quest for Omega*. Vintage, 2008.

19. Chakrabarty, Arnab. "Role of sensory input in structural plasticity of dendrites in adult neuronal networks." Ph.D. diss., der Fakultät für Biologie der Ludwig Maximilians Universität München, 2013. 88

20. Ciresan, Dan, Ueli Meier, Jonathan Masci, Luca M. Gambardella, and Jurgen Schmidhuber. "Flexible, high performance convolutional neural networks for image classification". *Proceedings of the Twenty-Second international joint conference on Artificial Intelligence*-Volume 2 (2011)1237–1242. 179

21. Cortes, Corinna and Vladimir Vapnik. "Support-vector networks." *Machine Learning* 20, no. 3 (1995): 273–297. DOI: 10.1007/BF00994018. 72

22. DeFelipe, Javier and Isabel Fariñas. "The pyramidal neuron of the cerebral cortex: morphological and chemical characteristics of the synaptic inputs." *Progress in Neurobiology* 39, no. 6 (1992): 563–607. DOI: 10.1016/0301-0082(92)90015-7. 49

22a. DeFelipe, Javier, Lidia Alonso-Nanclares, and Jon I. Arellano. "Microstructure of the neocortex: comparative aspects." *Journal of Neurocytology*, 31, No. 3–5 (2002): 299–316. DOI: 10.1023/A:1024130211265. 60

23. Diehl, Peter U. and Matthew Cook. "Unsupervised learning of digit recognition using spike-timing-dependent plasticity." *Frontiers in Computational Neuroscience* 9 (2015): 1–9. DOI: 10.3389/fncom.2015.00099. 18

24. Eliasmith, Chris, Terrence C. Stewart, Xuan Choo, Trevor Bekolay, Travis DeWolf, Charlie Tang, and Daniel Rasmussen. "A large-scale model of the functioning brain." *Science* 338, no. 6111 (2012): 1202–1205. DOI: 10.1126/science.1225266. 16, 18

25. Esser, Steven K., Paul A. Merolla, John V. Arthur, Andrew S. Cassidy, Rathinakumar Appuswamy, Alexander Andreopoulos, David J. Berg, Jeffrey L. McKinstry, Timothy, Melano, Davis R. Barch, Carmelo di Nolfo, Pallab Datta, Arnon Amir, Brian Taba, Myron D. Flickner, and Dharmendra S. Modha. "Convolutional Networks for Fast, Energy-Efficient Neuromorphic Computing." *Proceedings of the National Academy of Sciences* (2016): 201604850. DOI: 10.1073/pnas.1604850113. 44

26. Fauth, Michael, Florentin Wörgötter, and Christian Tetzlaff. "The formation of multi-synaptic connections by the interaction of synaptic and structural plasticity and their functional consequences." *PLoS Computational Biology* 11, no. 1 (2015). DOI: 10.1371/journal.pcbi.1004031. 63, 64

27. Felleman, Daniel J. and David C. Van Essen. "Distributed hierarchical processing in the primate cerebral cortex." *Cerebral Cortex* 1, no. 1 (1991): 1–47. DOI: 10.1093/cercor/1.1.1. 57, 59

28. Fidjeland, Andreas K., Etienne B. Roesch, Murray P. Shanahan, and Wayne Luk. "NeMo: A platform for neural modelling of spiking neurons using GPUs." *20th IEEE International Conference on Application-specific Systems, Architectures and Processors* (2009): 137–144. DOI: 10.1109/asap.2009.24. 16

29. Gautrais, Jacques and Simon Thorpe. "Rate coding versus temporal order coding: a theoretical approach." *BioSystems* 48 (1998): 57–65. DOI: 10.1016/S0303-2647(98)00050-1. 70

30. Gerstner, Wulfram and Werner M. Kistler. *Spiking Neuron Models: Single Neurons, Populations, Plasticity*, Cambridge University Press, 2002. DOI: 10.1017/CBO9780511815706. 51, 99

31. Gerstner, Wulfram, Richard Kempter, J. Leo van Hemmen, and Hermann Wagner. "A neuronal learning rule for sub-millisecond temporal coding." *Nature* 383, no. 6595 (1996): 76–78. DOI: 10.1038/383076a0. 103, 117

32. Green, Sara. "Can biological complexity be reverse engineered?" *Studies in History and Philosophy of Science Part C: Studies in History and Philosophy of Biological and Biomedical Sciences* 53 (2015): 73–83. DOI: 10.1016/j.shpsc.2015.03.008. 101

33. Guyonneau, Rudy, Rufin VanRullen, and Simon J. Thorpe. "Neurons tune to the earliest spikes through STDP." *Neural Computation* 17, no. 4 (2005): 859–879. DOI: 10.1162/0899766053429390. 69

34. Guyonneau, Rudy, Rufin VanRullen, and Simon J. Thorpe. "Temporal codes and sparse representations: A key to understanding rapid processing in the visual system." *Journal of Physiology-Paris* 98 (2004): 487–497. DOI: 10.1016/j.jphysparis.2005.09.004. 69

35. Han, Jie, Jianbo Gao, Pieter Jonker, Yan Qi, and Jose AB Fortes. "Toward hardware-redundant, fault-tolerant logic for nanoelectronics." *Design & Test of Computers, IEEE* 22, no. 4 (2005): 328–339. DOI: 10.1109/MDT.2005.97. 90

36. Hawkins, Jeff and Sandra Blakeslee. *On Intelligence.* Macmillan, 2007. 59

37. Hebb, Donald O. *The Organization of Behavior: A Neuropsychological Theory.* Psychology Press, (1949). 50

38. Hikosaka, Okihide, Yoriko Takikawa, and Reiko Kawagoe. "Role of the basal ganglia in the control of purposive saccadic eye movements." *Physiological Reviews* 80, no. 3 (2000): 953–978. 74

39. Hill, Sean L., Yun Wang, Imad Riachi, Felix Schürmann, and Henry Markram. Statistical connectivity provides a sufficient foundation for specific functional connectivity in neocortical neural microcircuits. *Proceedings of the National Academy of Sciences* 109, no. 42 (2012): E2885–E2894. DOI: 10.1073/pnas.1202128109. 58, 62, 63, 91

40. Hodgkin, Alan L. and Andrew F. Huxley. A quantitative description of membrane current and its application to conduction and excitation in nerve. *The Journal of Physiology* 117.4 (1952): 500. 97

41. Hopfield, John J. "Pattern recognition computation using action potential timing for stimulus representation." *Nature* 376 (1995): 33–36. DOI: 10.1038/376033a0. 103, 115, 117

42. Izhikevich, Eugene M. "Simple model of spiking neurons." *IEEE Transactions on Neural Networks* 14.6 (2003): 1569–1572. DOI: 10.1109/TNN.2003.820440. 93, 107

43. Johansson, Roland S. and Ingvars Birznieks. "First spikes in ensembles of human tactile afferents code complex spatial fingertip events." *Nature Neuroscience* 7, no. 2 (2004): 170–177. DOI: 10.1038/nn1177. 72, 86

44. Josephs, Mark B., Steven M. Nowick, and C. H. Kees Van Berkel. "Modeling and design of asynchronous circuits." *Proceedings of the IEEE* 87, no. 2 (1999): 234–242. DOI: 10.1109/5.740017. 28

45. Karnani, Mahesh M., Masakazu Agetsuma, and Rafael Yuste. "A blanket of inhibition: functional inferences from dense inhibitory connectivity." *Current Opinion in Neurobiology* 26 (2014): 96–102. DOI: 10.1016/j.conb.2013.12.015. 49, 87

46. Kermany, Einat, Asaf Gal, Vladimir Lyakhov, Ron Meir, Shimon Marom, and Danny Eytan. "Tradeoffs and constraints on neural representation in networks of cortical neurons." *The Journal of Neuroscience* 30.28 (2010): 9588–9596. DOI: 10.1523/JNEUROSCI.0661-10.2010. 72, 73, 86

47. Kheradpisheh, Saeed Reza, Mohammad Ganjtabesh, and Timothée Masquelier. "Bio-inspired unsupervised learning of visual features leads to robust invariant object recognition." *Neurocomputing* 205 (2016): 382–392. DOI: 10.1016/j.neucom.2016.04.029. 195, 196

48. Laurent, Gilles. "A systems perspective on early olfactory coding." *Science* 286, no. 5440 (1999): 723–728. DOI: 10.1126/science.286.5440.723. 52

49. LeCun, Yann, Léon Bottou, Yoshua Bengio, and Patrick Haffner. "Gradient-based learning applied to document recognition." *Proceedings of the IEEE* 86, no. 11 (1998): 2278–2324. DOI: 10.1109/5.726791. 73

50. Lee, Jungah, HyungGoo R. Kim, and Choongkil Lee. "Trial-to-trial variability of spike response of V1 and saccadic response time." *Journal of Neurophysiology* 104, no. 5 (2010): 2556–2572. DOI: 10.1152/jn.01040.2009. 73, 75, 86

51. Leung, Johahn, David Alais, and Simon Carlile. "Compression of auditory space during rapid head turns." *Proceedings of the National Academy of Sciences of the United States of America* 105, no. 17 (2008): 6492–6497. DOI: 10.1073/pnas.0710837105. 67

52. Lippmann, Richard. "An introduction to computing with neural nets," *ASSP Magazine*, IEEE 4.2 (1987): 4–22. DOI: 10.1109/MASSP.1987.1165576. 18

53. Maass, Wolfgang. "Networks of spiking neurons: the third generation of neural network models," *Neural Networks* 10.9 (1997): 1659–1671. DOI: 10.1016/S0893-6080(97)00011-7. 106

54. Maass, Wolfgang. "Noisy spiking neurons with temporal coding have more computational power than sigmoidal neurons." In *Advances in Neural Information Processing Systems* vol. 9, (1997): 211–213. 100

55. Maass, Wolfgang, Thomas Natschläger, and Henry Markram. "Real-time computing without stable states: A new framework for *Neural Computation* based on perturbations." *Neural Computation* 14.11 (2002): 2531–2560. DOI: 10.1162/089976602760407955.

56. Madhavan, Advait, Timothy Sherwood, and Dmitri Strukov. "Race logic: A hardware acceleration for dynamic programming algorithms." *ACM SIGARCH Computer Architecture News* 42, no. 3 (2014): 517–528. DOI: 10.1145/2678373.2665747. 45

57. Madhavan, Advait, Timothy Sherwood, and Dmitri Strukov. "Race logic: Abusing hardware race conditions to perform useful computation." *IEEE Micro* 35, no. 3 (2015): 48–57. DOI: 10.1109/MM.2015.43. 45

58. Madhavan, Advait, Timothy Sherwood, and Dmitri Strukov. "Energy efficient computation with asynchronous races." In *Design Automation Conference* (DAC), 2016 53rd ACM/EDAC/IEEE (2016): 1–6. DOI: 10.1145/2897937.2898019. 45

59. Magee, Jeffrey C., and Erik P. Cook. "Somatic EPSP amplitude is independent of synapse location in hippocampal pyramidal neurons." *Nature Neuroscience* 3 (2000): 895–903. DOI: 10.1038/78800. 64

60. Mainen, Zachary F., and Terrence J. Sejnowski. "Reliability of spike timing in neocortical neurons." *Science* 268, no. 5216 (1995): 1503–1506. DOI: 10.1126/science.7770778. 13, 71, 86

61. Maldonado, Pedro, Cecilia Babul, Wolf Singer, Eugenio Rodriguez, Denise Berger, and Sonja Grün. "Synchronization of neuronal responses in primary visual cortex of monkeys viewing natural images." *Journal of Neurophysiology* 100, no. 3 (2008): 1523–1532. DOI: 10.1152/jn.00076.2008. 68, 83

62. Markram, Henry, Joachim Lübke, Michael Frotscher, and Bert Sakmann. "Regulation of synaptic efficacy by coincidence of postsynaptic APs and EPSPs." *Science* 275, no. 5297 (1997): 213–215. DOI: 10.1126/science.275.5297.213. 50, 108

63. Markram, Henry, Joachim Lübke, Michael Frotscher, Arnd Roth, and Bert Sakmann. "Physiology and anatomy of synaptic connections between thick tufted pyramidal neurones in the developing rat neocortex." *The Journal of Physiology* 500, no. Pt 2 (1997): 409. 63

64. Markram, Henry, Maria Toledo-Rodriguez, Yun Wang, Anirudh Gupta, Gilad Silberberg, and Caizhi Wu. "Interneurons of the neocortical inhibitory system." *Nature Reviews Neuroscience* 5, no. 10 (2004): 793–807. DOI: 10.1038/nrn1519. 93

65. Masquelier, Timothée. "Relative spike time coding and STDP-based orientation selectivity in the early visual system in natural continuous and saccadic vision: a computational model." *Journal of Computational Neuroscience* 32, no. 3 (2012): 425–441. DOI: 10.1007/s10827-011-0361-9. 39, 147

66. Masquelier, Timothée, and Simon J. Thorpe. "Unsupervised learning of visual features through spike timing dependent plasticity." *PLoS Computational Biology* 3, no. 2 (2007): e31. DOI: 10.1371/journal.pcbi.0030031. 195, 196

67. McCulloch, Warren S., and Walter Pitts. "A logical calculus of the ideas immanent in nervous activity." *The Bulletin of Mathematical Biophysics* 5, no. 4 (1943): 115–133. DOI: 10.1007/BF02478259. 44

68. Mead, Carver. "Neuromorphic electronic systems." *Proceedings of the IEEE* 78, no. 10 (1990): 1629–1636. DOI: 10.1109/5.58356. 107

69. Mehta, Mayank R., Albert K. Lee, and Matthew A. Wilson. "Role of experience and oscillations in transforming a rate code into a temporal code." *Nature* 417, no. 6890 (2002): 741–746. DOI: 10.1038/nature00807. 67

70. Morrison, Abigail, Markus Diesmann, and Wulfram Gerstner. "Phenomenological models of synaptic plasticity based on spike timing." *Biological Cybernetics* 98, no. 6 (2008): 459–478. DOI: 10.1007/s00422-008-0233-1. 109

71. Mountcastle, Vernon B. "Modality and topographic properties of single neurons of cat's somatic sensory cortex." *Journal of Neurophysiology* 20, no. 4 (1957): 408–434. 59

72. Mountcastle, Vernon. "An organizing principle for cerebral function: The unit model and the distributed system." In *The Mindful Brain: Cortical Organization and the Group-selective Theory of Higher Brain Function.* edited by Edelman, Gerald M., and Vernon B. Mountcastle, eds. Massachusetts Institute of Technology Pr, 1978. 59

73. Mountcastle, Vernon B. "The columnar organization of the neocortex." *Brain* 120, no. 4 (1997): 701–722. DOI: 10.1093/brain/120.4.701. 59

74. Natschläger, Thomas, and Berthold Ruf. "Spatial and temporal pattern analysis via spiking neurons." *Network: Computation in Neural Systems* 9, no. 3 (1998): 319–332. DOI: 10.1088/0954-898X_9_3_003. 103, 117, 119

75. Natschläger, Thomas, and Berthold Ruf. "Pattern analysis with spiking neurons using delay coding." *Neurocomputing* 26 (1999): 463–469. DOI: 10.1016/S0925-2312(99)00052-1. 120

76. O'Connor, Peter, Daniel Neil, Shih-Chii Liu, Tobi Delbruck, and Michael Pfeiffer. "Real-time classification and sensor fusion with a spiking deep belief network." *Neuromorphic Engineering Systems and Applications* (2015): 1–10. 18

77. Perin, Rodrigo, Thomas K. Berger, and Henry Markram. "A synaptic organizing principle for cortical neuronal groups " *Proceedings of the National Academy of Sciences* 108, no. 13 (2011): 5419–5424. DOI: 10.1073/pnas.1016051108. 61

78. Peters, Alan, and Claire Sethares. "Myelinated axons and the pyramidal cell modules in monkey primary visual cortex." *Journal of Comparative Neurology* 365, no. 2 (1996): 232–255. DOI: 10.1002/(SICI)1096-9861(19960205)365:2<232::AID-CNE3>3.0.CO;2-6. 58

79. Petersen, Rasmus S., Stefano Panzeri, and Mathew E. Diamond. "Population coding of stimulus location in rat somatosensory cortex." *Neuron* 32.3 (2001): 503–514. DOI: 10.1016/S0896-6273(01)00481-0. 71, 86

80. Pierce, William H. "Interwoven redundant logic." *Journal of the Franklin Institute* 277, no. 1 (1964): 55–85. DOI: 10.1016/0016-0032(64)90039-0. 91

81. Wolfgang Maass, Henry Markram, and Marc-Oliver Gewaltig. "Liquid computing in a simplified model of cortical layer IV: Learning to balance a ball." *Artificial Neural Networks and Machine Learning*–ICANN 2012 (2012): 209–216.

82. Querlioz, Damien, Olivier Bichler, and Christian Gamrat. "Simulation of a memristor-based spiking neural network immune to device variations." *International Joint Conference on Neural Networks* (2011): 1775–1781. DOI: 10.1109/ijcnn.2011.6033439. 16, 18

83. Raghavachari, Sridhar, John E. Lisman, Michele Tully, Joseph R. Madsen, E. B. Bromfield, and Michael J. Kahana. "Theta oscillations in human cortex during a working-memory task: evidence for local generators." *Journal of Neurophysiology* 95, no. 3 (2006): 1630–1638. DOI: 10.1152/jn.00409.2005. 76

84. Reingold, Eyal M., and Dave M. Stampe. "Saccadic inhibition in voluntary and reflexive saccades." *Journal of Cognitive Neuroscience* 14, no. 3 (2002): 371–388. DOI: 10.1162/089892902317361903. 69, 83

85. Rumsey, Clifton C., and Larry F. Abbott. "Synaptic democracy in active dendrites." *Journal of Neurophysiology* 96, no. 5 (2006): 2307–2318. DOI: 10.1152/jn.00149.2006. 64

86. Scherer, Dominik, Andreas Müller, and Sven Behnke. "Evaluation of pooling operations in convolutional architectures for object recognition." *Artificial Neural Networks*–ICANN 2010, Springer Berlin Heidelberg, (2010): 92–101. 16, 152

87. Serre, Thomas, Lior Wolf, and Tomaso Poggio. "Object recognition with features inspired by visual cortex." *Conference on Computer Vision and Pattern Recognition* vol. 2 (2005): 994–1000. DOI: 10.1109/cvpr.2005.254. 195

88. Simard, Patrice, David Steinkraus, and John C. Platt. "Best practices for convolutional neural networks applied to visual document analysis." *ICDAR*, vol. 3, (2003): 958–962. DOI: 10.1109/icdar.2003.1227801. 179

89. Singer, Wolf. "Neocortical rhythms", in *Dynamic Coordination in the Brain: From Neurons to Mind*. Von der Malsburg, Christoph, William A. Phillips, and Wolf Singer, eds. MIT Press, 2010. DOI: 10.7551/mitpress/9780262014717.003.0011. 75, 84

90. Sjöström, P. Jesper, Ede A. Rancz, Arnd Roth, and Michael Häusser. "Dendritic excitability and synaptic plasticity." *Physiological Reviews* 88, no. 2 (2008): 769–840. DOI: 10.1152/physrev.00016.2007. 50

91. Smith, James E. "Efficient digital neurons for large scale cortical architectures." *Proceeding of the 41st Annual International Symposium on Computer Architecuture* (2014): 229–240. DOI: 10.1109/isca.2014.6853206. 99

92. Sohal, Vikaas S., Feng Zhang, Ofer Yizhar, and Karl Deisseroth. "Parvalbumin neurons and gamma rhythms enhance cortical circuit performance." *Nature* 459, no. 7247 (2009): 698–702. DOI: 10.1038/nature07991. 77

93. Spruston, Nelson. "Pyramidal neurons: dendritic structure and synaptic integration." *Nature Reviews Neuroscience* 9, no. 3 (2008): 206–221. DOI: 10.1038/nrn2286. 138

94. Stein, Richard B. "A theoretical analysis of neuronal variability." *Biophysical Journal* 5.2 (1965): 173–194. DOI: 10.1016/S0006-3495(65)86709-1. 105, 106

95. Stent, Gunther S. "A physiological mechanism for Hebb's postulate of learning." *Proceedings of the National Academy of Sciences* 70, no. 4 (1973): 997–1001. DOI: 10.1073/pnas.70.4.997. 50

96. Thorpe, Simon J. "Why connectionist models need spikes." Chapter 2: *Computational Modelling in Behavioural Neuroscience*, Vol. 2. Psychology Press, 2009. 69, 92, 128

97. Thorpe, Simon J. and Michel Imbert. "Biological constraints on connectionist modelling." *Connectionism in Perspective* (1989): 63–92. 67, 69, 86

98. Thorpe, Simon, Arnaud Delorme, and Rufin Van Rullen. "Spike-based strategies for rapid processing." *Neural Networks* 14.6-7 (2001): 715–725. 69

99. Tyron, J. C. Quadded logic, *Redundancy Techniques for Computing Systems*, pp. 205–228, Spartan, Washington, DC, 1962. 91

100. Uchida, Naoshige, and Zachary F. Mainen. "Speed and accuracy of olfactory discrimination in the rat." *Nature Neuroscience* 6, no. 11 (2003): 1224–1229. DOI: 10.1038/nn1142. 67

101. Unger, Stephen H. *Asynchronous Sequential Switching Circuits*, Wiley-Interscience, 1969. 28

102. Van Rullen, Rufin, and Simon J. Thorpe. "Rate coding versus temporal order coding: what the retinal ganglion cells tell the visual cortex." *Neural Computation* 13, no. 6 (2001): 1255–1283. DOI: 10.1162/08997660152002852. 147

103. Van Rullen, Rufin, Rudy Guyonneau, and Simon J. Thorpe. "Spike times make sense." *Trends in Neurosciences* 28.1 (2005): 1–4. DOI: 10.1016/j.tins.2004.10.010. 69

104. Vidal, Juan R., Maximilien Chaumon, J. Kevin O'Regan, and Catherine Tallon-Baudry. "Visual grouping and the focusing of attention induce gamma-band oscillations at different frequencies in human magnetoencephalogram signals." *Journal of Cognitive Neuroscience* 18, no. 11 (2006): 1850–1862. DOI: 10.1162/jocn.2006.18.11.1850. 77

105. Yuste, Rafael. "From the neuron doctrine to neural networks." *Nature Reviews Neuroscience* 16, no. 8 (2015): 487–497. DOI: 10.1038/nrn3962. 4

Author Biography

James E. Smith is Professor Emeritus in the Department of Electrical and Computer Engineering at the University of Wisconsin-Madison. He received his Ph.D. from the University of Illinois in 1976. He then joined the faculty of the University of Wisconsin-Madison, teaching and conducting research—first in fault-tolerant computing, then in computer architecture.

In 1979, he took a leave of absence to work for the Control Data Corporation in Arden Hills, MN, participating in the design of the CYBER 180/990. While at Control Data and after returning to the University of Wisconsin in 1981, he studied several aspects of high performance pipelined processors. This work included the development of dynamic history-based branch predictors, instruction issuing methods, and techniques for providing precise interrupts that are widely used today. From 1984–1989, he was principal architect and a logic designer for the ACA ZS-1, a scientific computer employing a dynamically scheduled, superscalar processor architecture. In 1989, Dr. Smith joined Cray Research and headed a small research team that participated in the development and analysis of future supercomputer architectures. This work focused on advanced vector processor implementations, high bandwidth memory systems, and interconnection networks.

In 1994, he re-joined the Department of ECE at the University of Wisconsin. His research interests were directed at new paradigms for exploiting instruction level parallelism. The virtual machine abstraction was used as a technique for providing high performance through co-design and tight coupling of hardware and software. In 2007, he retired from Wisconsin, and then conducted research in industry for four years, first at Google, then at Intel. He received the 1999 ACM/IEEE Eckert-Mauchly Award for contributions to computer architecture.

Currently, he is studying new neuron-based computing paradigms at home along the Clark Fork near Missoula, Montana.

Printed in the United States
by Baker & Taylor Publisher Services